Christmas 1998

In the Tracks of
Old Bluey

Sculpture of Nat Buchanan by Eddie Hackman
Photograph courtesy of Darrell Lewis

First published in 1997 by
Central Queensland University Press
PO Box 1615
Rockhampton, Queensland 4700

National Library of Australia
Cataloguing-in-Publication entry:

Buchanan, Bobbie (Roberta), 1943 - .
    In the Tracks of Old Bluey : The life story of Nat
    Buchanan

    Includes Index
    ISBN 1 875998 19 5

    1. Buchanan, Nathaniel, 1826-1901. 2. Explorers -
    Australia, Northern - Biography. 3. Australia, Northern -
    Discovery and exploration. I. Title.

    919.4290402092

Designed and typeset by Carlin Yarrow
in Times New Roman Monotype and Black Chancery.

Printed and Bound by
Watson Ferguson & Co, Brisbane

CQU Press would like to pay tribute to the magnificent painting on the front cover of this book. The painting is entitled "The Yarding of the Mob" by Sir Daryl Lindsay. CQU Press has made strenuous efforts through the National Trust of Victoria, through Dr Joseph Brown, a family friend of Sir Daryl, through Legend Press and McLaren & Co. to locate the private collection in which this painting is currently held. Although we have been unsuccessful in these efforts, we would like to acknowledge the unknown owner of this painting.

AQ

ARTS QUEENSLAND

Sponsored by the Queensland Office of Arts and Cultural Development

Contents

King Paraway

In the 80's Australia saw movements of cattle,
In the world's driest continent the drovers did battle
With nature, and thousands of hardmiles were spanned
As two hundred thousand were walked overland
Into Queensland, the Territory and the Kimberley runs
And in the forefront was one of the land's finest sons.
There are hundreds of drovers of whom we could sing
But everyone knows Nat Buchanan was King.

Chorus

Nat Buchanan, old Bluey, old Paraway
What would you think if you came back today?
It's not as romantic as in your time, old Nat,
Not many drovers and we're sad about that.
Fences and bitumen and road trains galore.
Oh they move cattle quicker but one thing is sure
Road trains go faster, but of drovers we sing
And everyone knows Nat Buchanan was King.

The bush blacks all called him old Paraway
You see him tomorrow, he left yesterday.
With thousands of cattle he keeps riding on
To nowhere, from somewhere, here he comes, now he's gone
With a bright green umbrella to shade the fierce sun.
On the Murranji, on the Murchison, on another new run,
Old Paraway's a man of whom desert tribes sing
And everyone knows Nat Buchanan was King.

Words and Music by Ted Egan

Acknowledgements

When I commenced making a list of all the people who assisted me in this project it became perfectly clear that 'No man is an island'. Friends, relatives and complete strangers have put their shoulders to the wheel to make *In the Tracks of Old Bluey* the document that it is. Thank you one and all for making it possible.

The support of the Northern Territory Government who generously provided money to complete the research through their History Awards scheme, is gratefully acknowledged.

Unfortunately I am unable to thank everyone personally. Many of the people who have been helpful are archive and library staff from around Australia - just doing their duty - but in a manner that has been patient and generous. My particular gratitude goes to the staff of the Northern Territory Reference Library, Northern Territory Archives, Northern Territory National Trust and the Northern Territory Place Names Committee.

My husband Chris Mobbs has been my mainstay during the several years taken to research and write the manuscript. He has supported me financially and emotionally and his absolute confidence in my ability to complete the task never wavered, although mine sometimes did. My request for a typewriter fell on deaf ears and for his insistence that I enter the age of technology and learn to use the computer, I will be eternally grateful. Not only is Chris responsible for producing the maps in the book but he has also been a wonderful sounding-board off which to bounce ideas. He has had the happy knack of listening to my obsessive ravings and offering some words of wisdom without hijacking the project.

Darrell Lewis of Darwin helped me from the outset and without him I would have put the task in the 'too hard' basket many times. He has always been enthusiastic and encouraging, not to mention hardworking. Darrell has assisted extensively with the research, provided photographs and set me in the right direction more times than I could mention. He has given unstintingly of his time and patience reading the rough manuscript, making corrections and giving the constructive and honest criticism required of a genuine friend. Thank you Darrell.

I have called on many people for information and assistance and would like to acknowledge their contribution. People who read and corrected the manuscript and gave their honest appraisal were Marilyn Pinkerton (Darwin), Rod Iddon (Toowoomba), and Cathie Clement and Peter Bridge (Perth). Marie Mahood

(Cattle Camp) has offered her time and knowledge generously, as have the following. Judith Hosier of Brisbane, Charlie Schultz (Proserpine), Eddie Hackman (Rockhampton), Ted Egan (Alice Springs), Olive Quilty, Colin Munro, Jill Bowen (Forestville), Leslie Cowper (Bowen Downs Station), Kate Deane (Bimbah Station), Mary and Glen Cameron (Mt Cornish Station, Qld), Pat and Peg Underwood (Katherine), Anna Underwood (Darwin), Patricia and John Westaway (Inverway Station), Terry and John Underwood (Riveren Station), Kathy Buchanan (Adelaide), Maggie Bruens (Adelaide), Beth Beckett (Rockhampton), Colin Beckett (Rockhampton), Brian Russell (Tamworth), Trin Truscett (Armidale), Tom Thornton (Tamworth), Peter Miller (Narrabri), Margaret Wass (Girilambone), Vern O'Brien (Darwin), Lorna Dudley (Humpty Doo), Peter Croker (Caloundra), Leonard Hill (Charters Towers), Reg Weston (Darwin). Last but by no means least are my daughter Elizabeth Risby who helped with the research and my sister Kate Buchanan who assisted with the Index.

My sincere thanks to the many others who have helped me along the way and have not been named. Your contribution has been invaluable.

List of Maps

Introduction

⁓⧚⧚⧚⧚⧚⧚⧚⧚⁓

Nat Buchanan was a legend in the Australian Outback long before his death in 1901. His name is still a byword in the bush and tales of his exploration and droving epics live on. Ted Egan has written songs about him and poets have paid him tribute in verse. Sculptor Eddie Hackman has immortalised Nat in bronze, and his achievements were recognised by the 1988 Bi-Centennial Committee as one of the 200 people who have made Australia great.

In a period which produced many outstanding bushmen, the quiet and unassuming Buchanan was without peer. He gained his reputation because of the enormous contribution he made to the settlement of the entire north of Australia. Many of his journeys were undocumented, but the ones that are disclose a list of firsts that are possibly unequalled.

With William Landsborough, Buchanan explored the Thomson River district in central western Queensland and in 1860 then was first to establish a station, Bowen Downs, in the area. He pioneered the stockroute from Bowen Downs up the Flinders River to the Gulf country and was first with cattle to Rocklands station on the Northern Territory/Queensland border. First also to cross the Barkly Tablelands from east to west and first to take a large herd of breeding cattle from Queensland to the Top End of the Northern Territory. For an encore,

In the Tracks of Old Bluey

The Life Story of Nat Buchanan

by

Bobbie Buchanan

Central Queensland
UNIVERSITY
PRESS

Buchanan created a droving record when he supervised the movement of 20,000 head of cattle over this route and became first to establish and stock a station in the Victoria River District of the Northern Territory - Wave Hill. Again, he was first with cattle to the East Kimberley in Western Australia to stock Ord River Station and first to cross the Murranji country with men and stock. First also to take cattle overland from the East Kimberley district to the Murchison River in Western Australia and last but not least, Buchanan was the first European to cross the Tanami Desert from Tennant Creek in the Northern Territory to Sturt Creek, Western Australia. Not only is this a formidable list but it also demonstrates that Buchanan was in no way a parochial explorer and pioneer - his work encompassed the entire north of Australia.

There were many pioneers who indulged in a bit of exploring essential to their own needs and survival, but none did this to anywhere near the extent of Buchanan. Among the many reasons for his success the following appear to have been the most influential. First, Buchanan was a master of bushcraft and these skills were enhanced by an unerring sense of direction, so extraordinary and keenly developed that it could only be considered a 'gift'. Although he took great risks, he didn't get lost or perish. Second, Nat was a natural leader of men and a great organiser. Third, his method of travelling light with only a few reliable, hand-picked men and the minimum of equipment gave his exploration trips more flexibility and he was better able to respond to prevailing circumstances. When it is all boiled down, he was doing what came naturally, what he was best suited for and what he best loved - looking for pastoral country.

Nat's journeys of exploration were done at his own expense and upon his own initiative. He responded to immediate needs and shouldered the costs and responsibilities personally. All his exploration, both recorded and unrecorded, was done in the cause of the advancement of pastoral settlement. It could be said that he did what he did out of self-interest and so is not worthy of the accolades accorded to more conventional explorers. However, they also answered the call to explore, in some respects, in response to the 'hip pocket nerve'. Buchanan never cost the country a penny nor did he receive any monetary reward for his endeavours. Certainly he took up valuable pastoral leases but often the best land that he discovered had already been taken up 'on the map', by city speculators.

Nat was his own man and valued his independence highly, so I suspect that he would not have taken kindly to instruction from city dwellers or government officials about how expeditions beyond settlement should be conducted. The red tape involved in organised exploring expeditions was not for him and he obviously knew it because, despite being well qualified, he only once offered his services to the government.

In an era when the Aborigines were maligned, misunderstood and often treated with great cruelty, Nat demonstrated a more enlightened outlook. Although always cautious in his dealings, he recognised the inherent bush skills of the Aboriginal people and their superior knowledge of the land. He was known for avoiding unnecessary confrontations by using conciliatory, innovative and peaceful methods, thereby defusing many potential conflicts. Carefully chosen and trusted Aboriginal men accompanied him on his exploration and droving trips, and they made a valuable contribution to the success of many of his missions. From all accounts the Aborigines he co-opted developed a great loyalty and respect for him.

From 1883 to 1901, Nat and his only child, Gordon (my grandfather), were closely associated and the pair travelled together on many occasions. Gordon wrote a first hand account of his father's life titled 'Packhorse and Waterhole', published in 1933 and again in 1934 by Angus and Robertson. In 1984, Hesperian Press published a facsimile edition of this book. As Nat Buchanan rarely recorded any of his journeys, Gordon's was the most comprehensive, contemporary account of his life. Besides his own first-hand knowledge of his father, he was able to call upon his mother Catherine Buchanan for much information. Hugh and Wattie Gordon, who were close associates of both Nat and Gordon, as well as a vast number of friends and acquaintances, also provided the material to piece together his father's life story.

Shortly before his death in 1943, Gordon Buchanan completed a manuscript titled 'Old Bluey'. This manuscript duplicated much of the material already published, but also added considerably to the fund of available knowledge and helped fill in some of the gaps in 'Packhorse and Waterhole'. Beside the vivid detailed descriptions of the methods Nat used for droving cattle, finding new stock routes, and journeys of exploration, new dimensions were added to his father's character which are worthy of recording. This manuscript was lost for many years but came to light, along with photographs, old letters and documents, in a moth-eaten cardboard carton in a dark corner of my parents' shed, after their deaths.

This unexpected find inspired me to amalgamate the information from the book and the unpublished manuscript to make a more comprehensive biography of Nat Buchanan. To this I have added information gained by extensive reading and research. Where possible I have authenticated dates and occurrences from primary sources in the hope that the finished product will not only be a good read, but also a reliable resource document. Where no information exists, I have constructed plausible hypotheses and have explained in the text my reasons for doing so.

A lot of information about Nat Buchanan has been misinterpreted, taken out of context or fabricated, with good intention, by a number of authors. I feel it is time the facts are told before any more misinformation is generated. Nat was a colourful, if enigmatic character whose story is quite remarkable and needs no exaggeration.

I have stolen shamelessly from my grandfather's writings because he was 'in the action' and speaks from experience whereas I can only imagine what it was like. Throughout the text I have quoted liberally from both *Packhorse and Waterhole* and *Old Bluey*. All quotations without specific endnote numbers are taken from these two works. Gordon Buchanan was a well educated 'bushy' and apart from some outdated terminology, especially in references to Aborigines, his work is a vivid portrait of his life and times and I believe for the purposes of historical record, that it should remain so. To avoid confusing the reader, I have differentiated Gordon Buchanan from the Gordon brothers by referring to him as 'Gordie', which was in fact the name used for him by his family and personal friends.

In *Packhorse and Waterhole* grandfather tended to get side-tracked into other interesting stories and although fascinating, this made the document disjointed and difficult to follow. Dates of important occurrences were sometimes missing or very vague. I have made a determined effort to set the biography down in chronological order and to record dates accurately where this is possible. Other noted explorers and outback figures have been placed in the context of the story and the timing of events, without going into their history in any depth. Most of them have been well documented elsewhere and as this is a biography of Nat Buchanan, I have tried to keep to that topic. In attempting to research the project thoroughly I have borne in mind the advice of a well-known historian who said that one has to be careful that the research doesn't overtake the project. Sometimes you have to go with what you have, if the document is ever to be published. No doubt there will be more information discovered, but with limited time and resources, what can be done has been done.

In some instances, especially in minor detail, there are discrepancies between *Old Bluey* and *Packhorse and Waterhole*. This is quite understandable as grandfather was old and ill when he wrote *Old Bluey*, and no doubt his memory sometimes betrayed him. On the other hand, he had been collecting information for years for the manuscript and he may have had new information from other sources which gave a different slant to the story. In these situations, and where unable to find another authoritative source, I have tried to present the most likely scenario in the context of other information I had. In some cases I have indicated in the text that the information is not verified. In many instances newspaper reports have been used to try and date incidents. Sometimes these reports appeared an indeterminate time after the actual events occurred, but they are often the only available record. Another invaluable source of detail have been the letters in my possession written to Gordon by his pioneering friends.

Nat's character is both simple and complex and not easy to describe. Despite all I have learned about him, many aspects of the man remain an enigma. For me this is part of the fascination that has kept me hooked. Because of my personal admiration for the man I hesitate to describe his character for fear of influencing the reader, but it seems I must.

Nat Buchanan was a confident, strong-willed and uniquely self-sufficient man of great integrity. People enjoyed talking to Nat, probably because he had the gift of being a good listener coupled with a quiet manner and good sense of humour. His tastes were modest, he was a wise judge of character and a committed peacemaker, which no doubt contributed to his organisational and leadership qualities. Consummate horseman, drover of renown and brilliant natural bushman sum up his practical skills. Always careful and prepared, he was a thinking man, ready for the unexpected. Slow to anger, he faced dangerous situations with a cool head, was scrupulously fair in his dealings with others and a loyal friend. Nat was a gentleman, his word was his bond and he expected the same in return. His friends and associates came from across the broad spectrum of the social milieu and he was respected equally by the highest and lowest in the land. On the Australian frontier where drinking, smoking and swearing were the norm he was one of the rare exceptions. His only addiction was to the Australian bush and in his later years he became a bit of a 'loner' and liked to 'go bush' for the peace and solitude it offered.

Nat was a quiet achiever but not much of a businessmen. He was always too busy getting on with the job in hand to do the 'paper work' and this was his downfall on Bowen Downs and perhaps on other occasions. Because he spent the majority of his time in unsettled areas he was required to entrust his affairs to others. One of his brothers, W.F. Buchanan, was a brilliant businessmen but Nat appears to have lacked the necessary will or 'killer instinct' to compete effectively in the cut and thrust of the world of commerce. Nat was definitely a quiet achiever.

I have described what may be seen by many as an exemplary character. This is not due entirely to bias on my part, but to lack of information to the contrary. A typical example of the opinion held of Nat by his contemporaries is that of William (Billy) Linklater. Billy often met Nat on the road and in mustering camps in the 1890's and had this to say of him: 'His willpower was indomitable, yet he was mild-mannered and of a most kindly disposition.'

Nat has been criticised for not taking his wife with him on his travels - did Stuart take his wife? Did Burke and Wills? How many drovers took their wives when they were blazing new tracks through the unknown? Surely it was more usual for women and children to remain behind until some sort of rudimentary accommodation was arranged.

Gordon described his parents' union in poetic terms but they must have experienced the stresses and strains suffered in most marriages. Perhaps the old saying 'Absence makes the heart grow fonder' was the secret ingredient that sustained them. The lifestyle Nat led prior to meeting Catherine would have precluded much association with women, yet he was discerning enough to recognise her as a kindred spirit and make her his bride. Catherine had come from a very close-knit, loving and supportive, pioneering family background and he formed lasting friendships with her brothers. Surely this would not have been so if Catherine's family thought Nat an unworthy partner. It is quite obvious that theirs was a 'love match' and although he was away for long periods the bonds remained strong throughout their thirty eight year marriage. In the complete absence of information to the contrary it appears that Nat found his perfect mate. From Gordie's writings it is obvious that he held his parents in high esteem. *Packhorse and Waterhole* and the *Old Bluey* manuscripts were expressions of his love, gratitude and admiration for them.

Gordie described his father's reasons for being a wanderer: 'The urge to explore and discover whatever the heart of Australia held seemed to be his ruling passion.'

Nat thrived on the challenge of the unknown and he loved the wide open spaces, the freedom of the outback. He had a discerning eye for selecting grazing country and he wanted to find suitable land to develop into a station worthy of his talents and experience. By the time he had taken up Wave Hill, however, his nomadic lifestyle was established and he could not rest anywhere for long. Even if Wave Hill had been the success story for him that he worked so tirelessly to achieve, he would still have been compelled to wander far beyond its borders.

In an era when Australians of Anglo-Celtic descent are encouraged to view their past as a shameful page in Australian history, I challenge the reader to acknowledge and applaud our cultural heritage. We can be proud of what is brave and good in our past whilst recognising and acknowledging that many deplorable deeds were committed in the name of settlement. Value systems and cultural norms have changed and it is unfair to judge our pioneers in hindsight from the comfort of our airconditioned living rooms. I believe that it was as much the Anglo-Celtic culture to colonise new land and farm it, as it was the Aboriginal culture to be hunters and gatherers.

I invite you to follow in the tracks of 'Old Bluey'.

Bobbie Buchanan.

1

The Crucible

1826 - 1859

⌐∾⟨∞⟩∾⌐

This is the story of my great-grandfather, Nathaniel 'Bluey' Buchanan, a unique Australian, who became an outback legend. In his later years he was known simply and affectionately as 'Old Bluey'. Beyond the black stump, to bushmen and drovers, he was famous for his skills as a pathfinder, explorer, drover and gentleman adventurer.

The origin of his nickname is obscure, but the most likely reason is that it was in recognition of the blue-grey colour of his hair and beard. Alternatively, it may have been inspired by the frequent mispronunciation of his name by the uneducated - 'Mr. Bluecannon'. The popular misconception that Nat was called 'Bluey' because he had red hair or 'humped a bluey' are certainly incorrect.

> *He never walked, he used to say he never worked but this was not true. He was always to be found astride a horse, borrowed, bought or bred, or else behind a pair or four in hand...We were always riding for the setting sun, or coming back to Sydney at long intervals, from Derby, Wyndham or Darwin by sea.* Special note

Most of Nat's early history has been lost but a few precious fragments survive. His story began in Wexford, Ireland, where he was born in 1826. His grandfather was Captain Andrew Buchanan of Mersheen, County Waterford while his father, Charles Henry Buchanan, was a Lieutenant of Her Majesty's 69th Regiment of Foot. Accompanied by his wife, Anne (nee White), and five sons, Charles arrived in New South Wales aboard the ship 'Statesman' in January 1837. Charles Henry was encouraged to come to Australia by his brother William, who had emigrated in 1822, and later written to his family in Ireland of the wealth of opportunities for advancement that existed in the new colony. Charles first settled with his family on the Hunter River, not far from his brother's property 'Mersheen'.

Nat was eleven years old when he arrived in Australia. The five boys in the family, from eldest to youngest, were Charles Henry, William Frederick, Francis (Frank), Nathaniel and Andrew. With no sisters, Nat grew up in a very male dominated environment. In 1839 the family moved to New England where Charles, in conjunction with his son William, took up the Rimbanda runs between Bendemeer and Uralla. Young Nat grew up close to the land and familiar with the bush. The boys were pressed into service to help develop and work Rimbanda, and possibly their lifestyle was fairly rigorously regimented by their military father.

It was an exciting time of new settlement on the New England frontier. The granite-studded and thickly forested country made the area an attractive hide-out for bushrangers and other rogues. When only sixteen and travelling alone in the Moonbi Ranges, Nat had a run-in with a bushranger named Wilson. He was on an errand for his father and riding along a narrow winding track when he was approached by another rider. Bearded and unkempt, the man had obviously been living rough in the bush. His horse had a dejected air and Nat could tell by its gait that it was footsore. When he drew nearer the man called to Nat to halt, aimed a pistol at his head, and demanded that he dismount. Nat was forced to remove his good boots and hand over his fresh horse Remaining silent and calm throughout the hold-up Nat obeyed the man's instructions quickly and to the letter. He must have been praying that Wilson would not discover his father's money in his pocket. Much to Nat's relief the bushranger was satisfied with the boots and horse. Although faced with a long walk home in dilapidated footwear leading a lame horse, Nat was probably relieved to have got off so lightly.

Nat's son Gordon said, 'he went to school in the normal way,' which gives us little insight into his education but suggests that it was appropriate for the times. Growing up close to nature in the picturesque, mostly virgin country of New England, Nat developed an extraordinary sense of direction. He grew into a great natural bushman and his son said that he, 'shared with the homing pigeon an innate sense of direction, which few others possessed.' His bushcraft and sense of direction were to become his hallmarks in future years.

In the late 1840s Nat went into partnership with his two brothers, Andrew and Frank, and took up the Bald Blair runs just north of Guyra. This property is still in existence today. It wasn't long, however, before the adventurous threesome heard of the California goldrush and with the enthusiasm and optimism of youth, joined the Forty-Niners. There were many boom and bust stories to come out of California at that time and the Buchanan story can only be described as the latter. Penniless, the brothers had to work their passage home on a windjammer. Nat was tall and lean and did not have a muscular physique, so his experience as a sailor left an indelible impression upon him. The ordeal of clinging, terrified, to the yardarm far above the heaving sea in inclement weather while wrestling with ropes and sails was not one of his happy recollections.

Once home the brothers were forced to relinquish Bald Blair due to mismanagement in their absence so Nat turned his attentions to droving stock to the goldfields of Victoria and overlanding cattle to South Australia. There must have been many interesting stories to come out of this period but they have been lost to us. Travelling so far afield with stock, Nat soon became recognised as a reliable drover and organiser. Two people who were associated with Nat at the time were his camp cook, David Buchanan, who later became a colourful if eccentric practitioner of the law, and one of his stockmen, who was eighteen year old, John Mackay.[1] Despite their age difference Nat and John became firm friends and soon they were both to make their mark in Queensland.

In the mid-1850s gold was found in New England and Nat's brother William claimed to have made a discovery, however others made similar claims. In 1849, while his brothers were adventuring in America he had gone prospecting in Gippsland and other fields. When he returned, he soon recognised the gold-bearing potential of the country closer to home. Gold fever was running high in New England by 1856 and 5,000 diggers descended upon Rocky River, near Uralla.[2] Charles Henry abandoned the running of Rimbanda to William and moved to the fields with his wife and and their sons Charles, Frank and Andrew. A lovely letter written by Nat's mother in 1859, to her sister-in-law, tells of her life living under canvas on the Rocky River.[3] Picturesque Rimbanda has survived into the 1990s.

By 1859 the movement to have Queensland proclaimed a Colony was in full swing. Looking for new opportunities, Nat, travelling by horse and dog cart, took the track north to Rockhampton. On the way he had two unexpected but providential meetings. The first was with William Landsborough's brother. Before moving to Queensland the Landsboroughs had been citizens of New England, so it is probable that Nat had already made their acquaintance or at least knew of them. The other, possibly more important, meeting was with John and Isabella Gordon.

The Gordons had emigrated from Scotland soon after their marriage. John was an experienced sheep man, having been an overseer for Major Innes on the New England properties of Farrucabad, Dundee and Mole. He was overseer at Mihi Creek near Uralla when he accepted the management of Ban Ban Station, near

Gayndah in Queensland. The long, slow trek from New England was undertaken by bullock wagon and riding horse. The three eldest children, Jessie, Catherine (Kate) and William George (Willie), were old enough to help their parents with the four younger children, Mary Ann, James Hugh and the twins Isabella Jane and Walter Robert (Wattie).

Kate was the 'apple of her father's eye', a natural horsewoman, and an accomplished rider. She enjoyed helping her father and he appears to have been the only person to call her 'Kitty.' He was fond of boasting, 'My boy Kitty is the most intelligent worker on Mihi Creek and my right hand man'.[4] Her devotion to her father and her skill as a horsewoman were put to the test near Ben Lomond. John Gordon was accidentally knocked unconscious and the closest doctor was at Glen Innes. Catherine immediately volunteered to make the lonely and dangerous ride to Stonehenge Station where there was a store and wayside inn. From here a messenger could be sent to the doctor at Glen Innes. It was a very dark night and the heavy rain made the unmade tracks and river crossings slippery and dangerous. Fearing for her father's life and ignoring her own safety, she rode at breakneck speed trusting in her horse's honesty and sure-footedness. At the Beardy River he refused to negotiate the boggy crossing. Undaunted she rode further upstream where the channel was not too wide and the banks were firm, and jumped him over. After completing the sixteen mile ride and delivering her message Kate declined rest and shelter and hurried back along the bush track to her family's campsite, where she promptly collapsed with exhaustion. Following attention by the doctor John Gordon recovered quickly from his concussion and in a few days the little party was able to continue.

The Gordons were nearing the end of their journey to Ban Ban when Nat rode up to their riverside camp. The tall, dark haired stranger was welcomed at the campfire, introduced to the family members and invited to share the evening meal. Catherine felt drawn to the bearded guest sharing the campfire and stole glances at him as she helped her mother prepare the food. Despite the confines of the strict etiquette of the times, cupid shot his magic darts.

Nat and John Gordon sat up late that night discussing mutual interests in New England, and speculating on the challenges that lay ahead for them in Queensland. Catherine, aroused by strange new feelings, lay awake in her swag listening to the talk around the campfire. That chance meeting on the bank of the Burnett River was the beginning of an enduring relationship of love and mutual devotion between Catherine and Nat, and the start of his long and happy association with the Gordon family.

References

Special note: All unattributed quotations are from Gordon Buchanan's *Packhorse and Waterhole*, 1933, Angus and Robertson, and *Old Bluey*, unpublished ms. The unpublished ms. is part of the private collection of Bobbie Buchanan.

1. John Mackay - Born in Scotland in 1839 he came to the New England District in 1855, where he lived on his parents' farm. Mackay worked a gold claim on Rocky River, overlanded stock with his friend Nat Buchanan and did some surveying for landowners before leading an expedition in search of grazing land in 1860. The party set out from Rockhampton but the exploration started from Marlborough in Queensland in March. In May they discovered the area of country between the Burdekin River and the watershed of the Isaac River which became known as the Mackay district. Mackay's expedition took place in the same period as Buchanan and Landsborough's, and they both used Marlborough Station as a jumping off place. See the *Australian Dictionary of Biography Vol. 5*, General Editor Douglas Pike, Melbourne University Press 1966.

2. *Australian Dictionary of Biography*, 1966, Melbourne University Press - William Frederick Buchanan.

3. Letter from Anne Buchanan (nee White) dated 1859, Buchanan Collection.

4. Although Gordon Buchanan always spelt his mother's name with a 'K' her birth certificate shows it spelt with a 'C' and the name on her gravestone is spelt Catherine. It has become popular for people to refer to Catherine as 'Kitty', but throughout *Packhorse and Waterhole* and *Old Bluey* Gordon always refers to his mother as Kate, the only exception being the mention that her father addressed her by the pet name 'Kitty'.

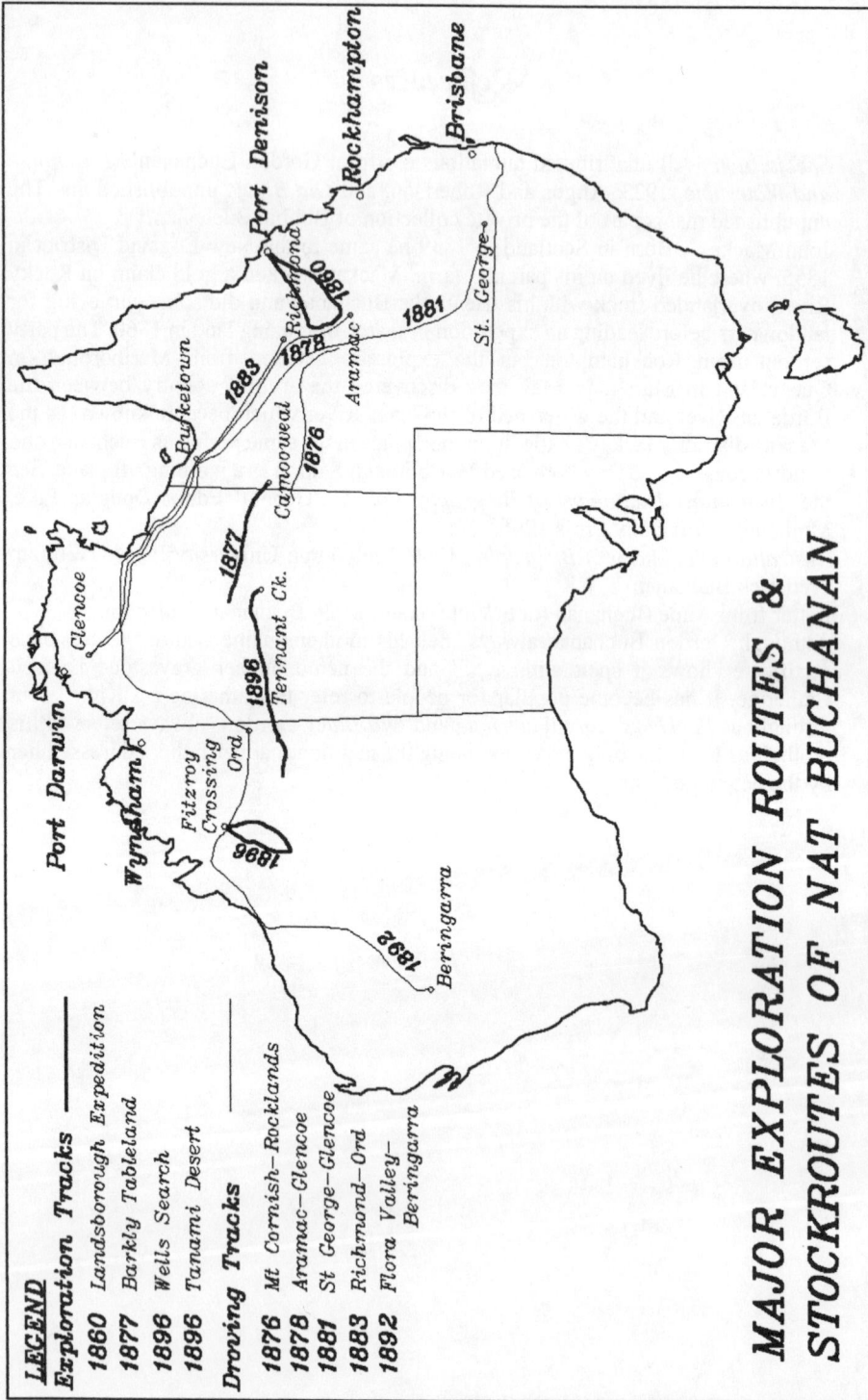

MAJOR EXPLORATION ROUTES & STOCKROUTES OF NAT BUCHANAN

LEGEND

Exploration Tracks ———

1860 Landsborough Expedition
1877 Barkly Tableland
1896 Wells Search
1896 Tanami Desert

Droving Tracks ———

1876 Mt Cornish–Rocklands
1878 Aramac–Glencoe
1881 St George–Glencoe
1883 Richmond–Ord
1892 Flora Valley–Beringarra

Map of Australia with Buchanan's droving and exploring routes marked

2. Lt. C. H. Buchanan - Nat's Father
Buchanan Collection

3. William Landsborough
John Oxley Library

2.

The Landsborough Expedition

1860

Rockhampton was established in 1856. It was a raw frontier town when Nat arrived and he took part, with considerable success, in the initial racemeetings conducted there. His first meeting with explorer and grazier, William Landsborough, took place in Rockhampton, maybe Landsborough's brother had given Nat a letter of introduction. Landsborough was already a well known Queensland identity and had an undeserved reputation as a 'professional' run hunter. Born in 1825 in Scotland, he migrated to Australia in 1841 and initially settled in New England with his two elder brothers. In 1854 he moved north with them to Queensland where, in association with John Ranken, they took up leases on the Kolan River between Gayndah and Gladstone. Landsborough started exploring in 1856 in the Broadsound and Isaac River area, where he named Mount Nebo and took up pastoral leases. In the following three years he explored the Comet River to its source, the Thomson River headwaters, and later in 1862, while leading a party from the Gulf of Carpentaria in search of Burke and Wills, he discovered the Barkly Tablelands.[1] A friend of Landsborough described him thus:

Landsborough's enterprise was entirely founded on his own self-reliance. He had neither Government aid or capitalists at his back when he achieved his success as an explorer. He was the very model of a pioneer - courageous, hardy, good humoured, and kindly. He was an excellent horseman, a most entertaining and, at times, eccentric companion, and he could starve with greater cheerfulness than any man I ever saw or heard of. But excellent fellow though he was, his very independence of character and success in exploring provoked much ill-will.[2]

Buchanan and Landsborough made a great team. Both bearded and with strong constitutions, Buchanan tall and lean and Landsborough big and burly. They were expert bushmen and horsemen and natural leaders of men, cool and courageous in danger and stoic and creative in adversity. Nat was modest and retiring by nature and not given to flamboyance or profanity. His dry humour complemented Landsborough who was a genial and gentle giant, and an uncomplaining travelling companion. Both shared an insatiable appetite for discovering new country. However they did differ in one very important respect.

Nat's sense of locality resembled the homing instinct of the plain pigeon, which, on open plains as wide and monotonous as the sea, returns to its nest in a tussock of grass with instinctive precision. Landsborough, on the other hand, was in this respect, only mediocre. Probably he could not get lost in a forty-acre paddock, but some of the paddocks now in Queensland would be enough to 'bush' him.

In 1860 Landsborough mounted an expedition to central western Queensland with Nat, Andrew Diehm and two Aborigines, Chucky and Tiger. This was a private exploration carried out at their own expense and was recorded by Landsborough in his diaries.[3] The first of Landsborough's diaries is missing so the starting point and exact date is not certain, probably Rockhampton or possibly from Landsborough's station, Glenprairie in the Broadsound region, early in February 1860. Their intention was to take a closer look at the Thomson River country discovered by Edmund Kennedy in 1847. Landsborough had seen this land very briefly in 1859, and he felt it held a lot of promise. Their route took them to Marlborough Station and then north west along creeks and rivers to near Mt. Douglas, across the Belyando, and on to Bully Creek. The party probably crossed the Great Divide north of what was later named Lake Buchanan, continued west to Torrens Creek and then to Towerhill Creek on the headwaters of the Thomson River.[4]

The land they travelled had experienced recent good seasons so the downs were well grassed and the ridges covered with saltbush. As they travelled they marked and ringed trees for future landmarks. The uniform nature of the downs country provided little chance for an overview. Being lean and wiry Nat was the one to shin up trees to get a better view of the surrounding landscape to report to Landsborough. He made sharp observations of the wild life - red-breasted parrots, top-knot pigeons and small white cockatoos, and descriptions of these made

Landsborough's dry diary notes more colourful. At one point they climbed a small table hill to get a better view of the country and here Nat discovered some tiny black berries. Eating them with relish he encouraged the rest of the party to do the same, saying they tasted as good as grapes. Landsborough noted in his diary that the country was studded with blackbutt, but open enough to allow sheep to graze through. The explorers became increasingly impressed by the grazing potential of the open downs.

Continuing their exploration they followed a creek coming from the east, later named Cornish Creek after Edward Cornish. Tracing this creek to its junction with the Thomson they found and named Landsborough Creek entering the main stream from the west. Following the Thomson south they discovered some soda springs and reedy waterholes close to the present site of Muttaburra. Pushing further south, they found another creek entering the Thomson from the east. Landsborough named it Aramak (Aramac) after his friend R. R. MacKenzie.[5] The discovery of these waterways radiating from the Thomson and their capacity to provide permanent water for stock reinforced their desire to secure the land for pastoral purposes.

This marked the southern extent of their exploration and from here they headed back to Cornish Creek exploring more country as they returned. Inadvertently they had pushed further south than their meagre rations would comfortably allow and this left them dangerously short of supplies. Their diet was monotonous, based on the staples of damper and tinned fish. On April 21st the shoeing tools were lost along the way and Nat retraced their steps in an unsuccessful search. Without the tools their supply of horseshoes was quite useless, so they buried them. This, plus the reduced supplies, lightened the load on the tiring horses.

The fact that the horses could no longer be shod probably contributed to Nat's favourite horse, Pickwick, developing lameness from a hoof injury. Nat reluctantly set him free to fend for himself. Pickwick was highly prized by Nat because he was a big useful chestnut horse with the easy comfortable gait of a good hack. Although he hoped to find him again when he returned to that country he knew the chances were remote.

Aborigines were observed in the area but they kept clear of the whites. Some were seen camped on a large lagoon on Cornish Creek, but they swam away to the other side at the approach of the party. Landsborough decided to set up camp by the lagoon where the plain was high and there was plenty of water. He considered the campsite to be an ideal situation for a future station homestead.

On the 25th of April, they crossed the Drummond Range and about thirty miles away to the east saw the Peak Range. Quite some time was lost because of ongoing problems with horses getting lost or knocked up. One horse managed to get into Campbells Creek but could not get out, so this required another day trying to extricate it. As the horses were nearly done in and probably foot-sore, they took periodic stops to rest them and this slowed their progress further.

On the 2nd May Landsborough made his first reference to the 'long necks' of the tucker bags. The men tried to shoot game with their carbines, but didn't have much luck although Nat shot a couple of bush turkeys which provided fresh meat for a couple of meals, and Andrew Diehm shot a duck. A few days later only four pounds of sugar and about the same amount of flour were all the provisions that remained and this made it imperative that they find a station. The flour, which had previously been considered unfit to eat because of lumps and weevils, was now put to good use. Empty bullock-hide bags which had been used to carry provisions were cut up and boiled, to produce an unappetising jelly, which the men ate. Landsborough, typically stoic, said it was, 'An excellent addition to our short allowance of provisions.'[6]

The weary and half starved little party made camp for the last time on the 11th May and marked a tree L/XCIV, recording the number of days since they left Marlborough Station. Nearly three and a half months had elapsed since their departure. Landsborough's homestead on his property Glenprairie, twenty-four miles from Taloombah, must have been a welcome sight to the travel-weary party, all of whom were skinny and suffering from the privations of their long journey. Landsborough's emaciated body was covered with boils, but he never complained. Home at last, with nourishing food to repair their health, they could begin to reflect upon the fine grazing land they had found and explored, and to work out how to make it theirs.

References

1. *Australian Dictionary of Biography* and E. Favenc, *The History of Australian Exploration 1788 - 1888*, Syd 1888.
2. Ernest Favenc, op. cit.
3. *Diary of W. Landsborough* 11/4/1860-11/5/1860, Landsborough Collection OM 69/ 30, J. Oxley Library, Brisbane.
4. The first diary of this expedition is missing and entries commence in the second diary on Wednesday April 11th from camp 64. The exact point of departure and the exact route taken is therefore unknown. It does seem likely, however, that Landsborough would have taken a route similar to that of his 1859 journey to the central west and this is recorded, so I have used it to indicate the approximate route taken for the 1860 journey.
5. W. Landsborough, *Journal of Queensland Expedition from Albert River in search of Burke and Wills*. 1862. Landsborough Collection OM 69/17, J. Oxley Library Brisbane.
6. In *Packhorse and Waterhole*, G. Buchanan reports Buchanan taking a trip with Landsborough to the central west in 1859 where the pair nearly starved and were rescued by John MacKay. In Landsborough's notebooks recording this trip there is no mention of Buchanan and so I have not included this episode.

The Landsborough Expedition

3.

The Bowen Downs Saga

1861-1866

———⋙⋘———

1861 saw the genesis of Bowen Downs, the first property established in the Longreach area of central western Queensland and one of its greatest early grazing enterprises.[1] Nat and Landsborough set out to secure leases over the runs and begin the task of developing the wilderness into a profitable grazing venture.

In January/February 1861 Nat returned to the downs country, this time in the company of Edward Cornish, a mutual friend of Landsborough and Robert Morehead of the Scottish Australian Company. Cornish was experienced in pastoral matters and was keen to get a first-hand look at the land. With Nat as his guide he rode over the country and was most impressed, concluding that it showed great potential for wool-growing. Before returning to more settled areas Nat and Cornish decided to make an exploratory trip to the west of Bowen Downs. Well equipped and with a black guide and a string of ten to twelve horses they set out to discover how far the downs country extended. Although they had all the land they could stock, their curiosity led them on.

They reached the east and west Darr and Vergemont Creeks in 150-180 miles, then continued on to the Diamantina. Here, to their great surprise and puzzlement, fresh camel dung and camel tracks were discovered heading in a northerly direction. The Aboriginal tracker was set on the trail which led them to a recently vacated camp. Because there were obvious indications of the presence of white men, the tracker continued on the trail. The tracks were followed for some miles in the hope of overtaking the outfit, but eventually the chase was abandoned when it led them far off course and they could not risk running out of provisions. The two scouters returned to the Landsborough Runs by a more northerly route, through the area around where Winton now stands.

The expedition they had so narrowly missed was none other than that of Burke and Wills, which set out from the Cooper Creek depot on 16th December 1860 and struck the Gulf on 11th February 1861.[2] These ill-informed, and inadequately prepared explorers plunged wildly and impatiently into the interior in an effort to reach the north coast. The expedition was among the most costly, and next to Leichhardt's final journey, the most tragic and the least beneficial of any in Australian history. However, the names of Burke and Wills are almost as well known today as they were then, while others, less wasteful of resources and successful in their missions have long since been forgotten.

In the meantime Landsborough was busy trying to secure the runs. The cost of a licence was ten shillings per square mile. As the runs totalled 1,500 square miles this was a considerable sum and Landsborough only secured them for twelve months. For a lease to be finally granted the land had to be stocked at a minimum of five head of cattle or twenty five head of sheep per square mile. On top of this were the costs of establishment which included - wages, equipment, provisions and stock - altogether a daunting sum.

To raise money Landsborough sold all the pastoral leases that he owned except Glenprairie.[3] Edward Cornish bought Landsborough's huge Fort Cooper runs for 16,000 pounds and in July 1861 installed A. Kemmis as manager.[4] Landsborough failed to raise sufficient money and for the venture to succeed a financial partner or partners had to be found to share the burden. A Sydney businessman, William Walker, bought a half share in the runs but later got cold feet and offered his share to the Scottish Australian Company. The Australian manager of this company, Robert Morehead, with encouragement and advice from Edward Cornish, had the foresight to recommend the investment to the parent company. The proposal was well received, thanks to Morehead's persuasive skills and his conviction that the company would reap rich rewards, given time and the initial expenditure. Edward Cornish was appointed General Superintendent and Nat Buchanan Resident Manager.

Morehead set up a five year partnership agreement which was to be renegotiated in October 1866.[5] The other shareholders were Landsborough and Cornish, and they agreed to advance Nat a one eighth-share in order to secure his services in establishing the property.

On November 21st 1861, Nat sailed from Sydney aboard the vessel 'Amity', bound for Port Denison (now Bowen) on his first assignment as resident manager of the Landsborough Runs Operation, as the project was initially called. His job was to open up a road suitable for transporting supplies and droving stock from Port Denison to the Western Downs, over 350 miles of largely unexplored inhospitable country, populated by sometimes hostile Aboriginal tribes. This road had to cross forests and plains, creeks and rivers, and the biggest obstacle of all, the Great Dividing Range. With four men, it took Nat several months to blaze the track which was to become the station's lifeline for the next few years. From Port Denison the track went to the last outpost of civilisation, Fort Cooper, then on to the Suttor River via Suttor Creek, and west to the Belyando River. The trail had to be clearly marked for others to follow, suitable places to cross water courses had to be found and where necessary cuttings made in their banks for the passage of horse drawn vehicles. Once across the Belyando the travel was easier as they followed Bully Creek towards the formidable barrier of the Great Divide.

For a while there was some delay while Nat made daily explorations of different routes before the most suitable path was found. Passing through the range Nat discovered a large expanse of water directly in their path. He was delighted because it was well placed for watering travelling men and stock, but to his chagrin the lake turned out to be salt and quite useless for his purpose. He must have been sorely disappointed when he named it Lake Buchanan.[6] Pressing on, they came to Torrens Creek which they ran down to Cornish Creek and their destination. Mission accomplished, Nat returned to the coast and sailed for Sydney to tender his report to Morehead and Cornish. After a brief stay attending to business matters he returned to Rockhampton on the steamer 'Boomerang' on the 25th May.

At the half yearly meeting of the Scottish Australian Investment Company held on August 1st. 1862, it was reported that the leases for the Landsborough runs had been acquired from the government and that a good road from the property to Port Denison had been established by the resident manager.[7] There was some urgency to meet the stocking requirements set down by the government by the appointed time so it was decided initially to stock the land with cattle.

Nat, back on the downs country, prepared for the arrival of a draft of 5,000 head of cattle which was being gathered at Fort Cooper. Nat was pilot for this first movement of stock along the new track to the Landsborough Runs. The droving party consisted of Walter Kerr in charge, assisted by Maurice Donohue plus four Aboriginal stockmen, three Aboriginal women and a boy.

The Aborigines inhabiting the surrounding country had a reputation for protecting their tribal lands and the drovers anticipated that they might have trouble. Nat's practice was to avoid conflict where possible and he usually achieved this by gaining the Aborigines' confidence or outwitting them. Unfortunately while he was away from camp scouting the country up ahead, a party of war-painted tribesmen appeared. Although no overt act of hostility was committed by the Aborigines, Kerr panicked and opened fire.[8] This inflamed the situation and reluctantly the rest of the party was obliged to join in. When Nat returned to the camp he was displeased and disappointed that no peaceful attempts had been made to prevent the incident. He was aware that injustices of this nature could threaten future peaceful relations with the Aborigines of the area and so endanger the lives of those using the route in future.

Stock losses were few, but when a dry stretch of about forty five miles was encountered beyond Bully Creek Nat decided to leave 1,500 cattle on that creek to be collected later. The first cattle arrived on the Thomson River in October 1862. After these were spread evenly across the runs to comply with stocking regulations, the cattle on Bully Creek were mustered and brought in.[9] Without fences the job of keeping the stock spread out over the runs was constant. Even well handled cattle, trained to stay on the run, roamed miles from the homestead keeping Nat and his stockmen constantly in the saddle. This story had a sad postscript when Maurice Donohue died of an unknown cause soon after reaching Bowen Downs. He was the first person buried on the property.

The second big draft of cattle was delivered to the station in May 1863. Drovers Hill and Bloxam, with two stockmen, Burkett and Best, set out from the Narran River in the Walgett District of N.S.W. in charge of 3,000 cattle. Their route was via the Maranoa River to the head of the Warrego, crossing to the head of the Barcoo. Hill and Bloxam were among the first cattle drovers to attempt that route, which at that time lay beyond the outskirts of civilisation. The season was very dry and the drovers had a hard time. A massive 1,000 head were lost along the way from thirst and starvation.[10]

By the end of 1863 the runs carried the minimum stock to secure the leases. This was a year of severe drought which denuded the usually abundant pastures and dried up the waterholes. It is thought that the initial hub of the property, a cluster of bark huts, was situated on the Thomson River south of Muttaburra, on the bank above a deep waterhole. A short time later, for some unknown reason the homestead site was moved to rising ground above a permanent waterhole on Cornish Creek. The level of the water was reduced to a meagre two feet in the 1863 drought.[11]

Before the days of independent carriers a lot of extra work and cost was created getting supplies from the coast to the station. This made it necessary to keep drays and working bullocks, and required valuable manpower to be released from stock work to make periodic trips to Port Denison. Bullocks were chosen as the draft animals because they had an advantage when the going was boggy and could pull heavy loads on the track, where only poor quality feed was available.

Edward Cornish visited the station again in 1863. As General Superintendent he was required to report back to the Company and of course he had a personal interest in the development of the runs. Cornish broke the news to Nat that he intended to purchase 25,000 young, well bred sheep and that they would soon be on the road for the station. This mob of sheep probably arrived late in 1863, because Nat was reported to have finished the first shearing in February 1864.[12]

The arrival of the sheep posed a different set of problems for Nat. The absence of fencing meant that shepherds had to be employed to tend the sheep out on the runs. Shepherds required some rudimentary accommodation on site and had to have regular rations of flour, tea, sugar and meat delivered to their far flung pastures. The guardian of the sheep led a difficult life. Besides flocks being harassed and scattered by dingoes and the lambs cruelly mauled, the lonely shepherds and their sheep were also the targets of Aboriginal attacks. It is not surprising that there was a high turnover of stockmen and shepherds which created ongoing problems for Nat. To attract good shepherds to work in such isolated, dangerous and primitive conditions high wages had to be offered as compensation and thirty five shillings per week was the going rate. Higher wages did attract men to the jobs, but they didn't stay long. Once they had earned enough money they left for more populated areas where they could put their cheques to good use, often in the grog shops. The situation was so bad that Nat decided that on his next trip to Sydney he would call at Brisbane and attempt to recruit shepherds for the Runs from newly arrived German immigrants.[13]

Since Nat and Kate's first meeting on the Burnett River in 1859, Nat had been a regular, if infrequent visitor of the Gordon family at Ban Ban Station. During busy years spent exploring and establishing Bowen Downs, Nat had taken each rare opportunity to call on Kate. He was very popular with the family and on these brief visits Kate had to compete with her father and her brothers for Nat's attention. Their moments alone were few and precious. Catherine was a pretty girl with serene brown eyes set in a fair-skinned oval face crowned with long and lustrous brown hair. The big difference in their ages didn't appear to affect their attraction to each other and it wasn't long before they had an informal understanding. Despite impediments of time, distance and lack of privacy, the couple managed to conduct their courtship. When Nat felt that he had a secure future to offer Kate he invited her to the privacy of the stables, using the excuse that he wanted to get her opinion of a new horse he had bought. They emerged engaged. The horse probably didn't receive any attention at all.

By the age of thirty six Nat was a Justice of the Peace in the Mitchell District and a well known identity in Queensland.[14] His reputation as a bushman was second to none, he was a horseman and drover of renown, and more recently, a grazier. Apart from their mutual attraction and a common interest in horses, the pair shared a New England background and the Presbyterian faith. They complemented each other like opposite sides of the same coin. Kate was a first generation Australian, born at Port Macquarie in 1842. Like Nat, she had grown up on the frontiers of civilisation and was no stranger to pioneering life. Beneath her soft voice, gentle manner and compassionate heart lay courage and tenacity.

Her greatest strength was her deep Christian belief which upheld her in later years during times of hardship, loneliness and uncertainty. Catherine had the selfless spirit of a true woman pioneer, and was the perfect mate for 'Bluey' Buchanan.

On the 19th August 1863, Nathaniel Buchanan married Catherine McDonald Gordon. The ceremony was celebrated at the Royal Hotel, Maryborough.[13] The entire Gordon family drove from Ban Ban to Maryborough for the happy occasion and John Gordon gave the bride away. Catherine's sister Annie and brother William were the attendants and witnesses. In Gordie's words:

> *We are told that the ideal union is achieved by the mating of two "opposites," or, at least, two dissimilar personalities; but this is not always true. In the mating of Katherine Gordon and Nat Buchanan the two personalities were rare spirits, perhaps, but they were anything but dissimilar; and the union was in every way happy and ideal. They were both pioneers, looking forward to the far future and ignoring the vicissitudes, however harsh, of the mere present; in this each upheld the other, each strengthened the other's heart to make little of mere ease and comfort, to sacrifice all for the future.*

Even for those times the honeymoon was unorthodox. It commenced with a short sea voyage to Port Denison. There were no port facilities in those days so the horses had to be swum ashore and the newly weds crossed from ship to shore in a rowing boat. Keen to get on the road, they set out immediately by buggy and pair on the 360 mile journey to Bowen Downs. Owing to a particularly dry season and the distance between waterholes, Nat was obliged to travel long stages at a fast pace.

On the outskirts of Port Denison the man driving the loose horses behind the buggy stopped for a drink at the last grog shop. Some drunks from the shanty noticed the unescorted buggy pass and fancied it might be easy pickings. They galloped after the buggy in an attempt to hold it up but when they got close they found Nat had his carbine ready and their bravado deserted them. Later that evening the repentant man caught up. Next day the pace was again brisk and he had a hard time keeping up with the spare horses. He made the mistake of thinking Kate would be sympathetic to his cause and described Nat as a madman. Kate bristled and quickly put him in his place, saying that this country needed more madmen like her husband.

There were plenty of signs of Aborigines around so each night a guard was mounted and Kate took her turn at watch along with the men. They were not attacked, although there was evidence that some grass fires had been set which were intended to burn out their camps.

Although only twenty one, Kate was familiar with the pioneering way of life, but as the first and only woman on Bowen Downs she faced new challenges. Her nearest neighbours were at Aramac Station fifty miles away to the south east, so the isolation was greater than she had ever experienced before. John Rule and Dyson Lacey pioneered Aramac Station early in 1863, when they brought sheep and settled on Aramac Creek. With Nat away, often for extended periods, she had to always be alert and prepared to cope with dangerous and unexpected situations that arose on the property. The Aborigines were becoming increasingly bold and the little outpost was often under threat of attack. Kate kept busy making her spartan accommodation more homely, preparing meals, attending the men when they were sick or injured, and trying to establish a vegetable garden to augment their diet. To the lonely men on Bowen Downs she must have brought a breath of civilisation and gentleness.

The country was in the grip of a serious drought and financial losses caused by ever increasing depredations by the Aborigines, plus the high cost of wages and supplies were all contributing to the station's growing debt. Nat was kept busy out on the runs, overseeing the stock, collecting scattered and wounded animals and attempting to protect the herds from Aboriginal attacks.

Kate had no sooner settled into her new home when she became aware that she was pregnant. Because of the primitive and often dangerous living conditions and the absence of another white woman or doctor in the area, Nat insisted that Kate return to Ban Ban for the birth. With one man to tend the horses, the pair set out on the long and dangerous journey to Rockhampton.

As it was December and the wet season was nearly upon them he set a scorching pace in an attempt to beat the storms. A few early showers fell, but this didn't slow them down until they reached the swollen Belyando River which was running a banker. Kate was prepared to swim across, but Nat made a raft of packsaddles, saplings and canvas to float her over while he swam in front with a tow line. More trips followed with the provisions and gear, and finally, the most difficult task, getting the buggy across. Alone on the far bank with a pistol ready for protection, Kate surveyed a vista of comparative calm. The only sounds were those of the gurgling river and the bumping of floating logs and debris as the turbid swirling waters of the flood-swollen stream rolled to the sea. The bush all around held a sinister menace. Here in the leafy shadows lurked the stealthy Aborigines, grimly evidenced by the dun-coloured smoke and crackling caused by their attempts to fire the undergrowth. Luckily it was too green to burn readily and the fire didn't spread. Kate remained alone and unprotected while the men on the opposite bank often moved out of sight to get timber for the raft and attend to the horses which were grazing some distance away. Despite the crackling undergrowth and drifting smoke she stood firm, ready to guard herself and their belongings until Nat and the man returned. The tribes of the Belyando had a ferocious reputation, so a watch was set and they slept nervously that night.

At a camp further on, Kate woke at dawn to the sound of Aborigines in the distance. Alarmed, she left her hammock only to find that Nat and his man were not in their swags or anywhere around the camp. Then to her great relief she was reassured by the sight of the man carrying up a billy of water from the creek and Nat, some distance away, bringing in the horses.

On another occasion Nat took a bridle and a pistol and went out to fetch the horses. Once out of sight of the camp he became aware of an Aborigine stationed between himself and the horses and on looking back saw his path back to camp blocked by another armed Aborigine. They shook their spears in a threatening manner and made warlike gestures which left Nat in no doubt of their intentions. Nat instinctively assumed a peaceful attitude, smiled at the man between himself and the horses and put his bridle down on the ground in front of him. This simple act of a genuine peacemaker completely defused the situation. Nat left the bridle where it lay and approached the horses and caught one using his belt and led it back to the bridle. The Aborigines were most surprised to see what use he then made of the bridle. Lesser men would probably have shot first and asked questions later.

Very soon they reached some outlying stations where they were welcomed and were able to enjoy some rest and comfort. Once they reached Marlborough Station they found that they only had one day to travel the seventy miles to Rockhampton, if they were to catch the S.S. *Queensland* on New Year's Day.[16] Departing before dawn and setting a cracking pace they arrived in Rockhampton after dark but in time to meet their deadline. Next day they sailed south taking the buggy and four horses with them.

The pair travelled to Sydney where Nat reported to Cornish, who told him to expect another 25,000 sheep on the station in twelve months. After a brief three day Sydney stopover Nat and Kate were once more on board ship heading up the coast to Brisbane where he intended to recruit German immigrants as shepherds. It was imperative that Nat not lose any time because he had to be back on Bowen Downs for the first shearing, which was soon to commence. The ship sailed via Maryborough where it is assumed that Kate disembarked, while Nat went on to Rockhampton. The first wool clip was much better than anticipated, amounting to 140 bales of high quality wool. That year there was heavy winter rain causing floods that washed away sheep and prevented shipment of the wool to the coast. When rain fell, teams got bogged down on the blacksoil plains and were often delayed for weeks.

Gordon was born at Ban Ban Station on the 29th May 1864.[17] He was to be Nat and Catherine's only child. Why this was so is a mystery. Mother and baby remained at Ban Ban for some time gathering strength and waiting for Nat to escort them home.

Bowen Downs became an important stopover for travellers en route to the Gulf Country. Sometime in this same year Nat made the acquaintance of Donald McIntyre when he passed through Bowen Downs on his way north to form a property for his brother Duncan. Duncan, who was a good bushman, had made an exploratory trip to the Gulf and had discovered what he thought to be old horses which could have belonged to Leichhardt and also some trees blazed 'L'. Sceptics believed that Landsborough was probably responsible for these but he managed to find backers who paid him 1,500 pounds to lead a search for Leichhardt. The expedition turned out to be a debacle from beginning to end, culminating in Duncan's death from fever. He was buried at Dalgonally, the station established in his absence by Donald. It lies north west of the present town of Julia Creek on the old droving route to the Gulf up the Flinders River. Donald McIntyre and Nat Buchanan formed a lifelong friendship.

At this time there was a race on to develop the Gulf Country. It had been a subject of speculation since 1862, when William Landsborough and others had made favourable reports about the Albert and Nicholson River country. Cornish and Morehead had selected an area between the two rivers, situated on a running stream called Beames Brook, which formed the head of the Albert River, and they referred to it as 'The Paddock'. The Scottish Australian Company made plans to stock this country before anyone else. With this in mind, Cornish told Nat to have cattle, droving plant and supplies ready to move up to the Gulf as soon as he heard news of other settlers making their way north with stock.[18] Because Bowen Downs was on the direct route to the Gulf, it was the ideal jumping-off place. Cornish, who had reservations about the healthiness of the Gulf country for men and sheep, was enthusiastic about its suitability for cattle and the opportunity it provided as a much needed outlet for Bowen Downs stock.

About mid-1864 news was received by the Company that a syndicate headed by John Robertson had stock on the road for 'The Paddock' and intended racing the Company for it. This was the signal for Nat to leave immediately for the Gulf with his prepared stock and plant. On the race to stock 'The Paddock' Nat blazed the first stock route down the Flinders River to the Gulf. This became the accepted route for the movement of cattle to that country and later formed part of the route to the Northern Territory via the Gulf Track.

Despite the Aborigines in the area having an aggressive reputation, Nat didn't encounter any serious trouble. The race to the Gulf was won by the Bowen Downs men and cattle when they arrived at Beames Brook in October 1864, well ahead of their competitors. Donald McGlashen, with three men and six months rations remained on 'The Paddock' to supervise the stock.[19] Returning to Bowen Downs, Nat prepared three more drafts of 2,000 head for the journey north. It had been very dry and in December the Aborigines set grass fires which destroyed a large number of sheep and scattered the cattle to all corners of the runs.

The Company was so enthusiastic about the Gulf Country at this stage that they seriously considered disposing of the drought-affected, debt-ridden Bowen Downs. Cornish, however, never lost faith in the downs country and strongly supported retaining the holding.

The distance and the high risk attached to droving stock across a belt of poisonous country to Port Denison prompted the Company to find an alternative outlet for their stock. They proposed a boiling down works on the Gulf which could reduce stock to tallow for export. Cornish was responsible for setting this up and it was first managed in 1866, by E.R. (Rowley) Edkins and his brother Henry. The works were not in production long before it was decided to take the Bowen Downs cattle to Cleveland Bay (Townsville) for boiling down. At the end of 1867, Rowley Edkins was appointed manager of Beames Brook Station and took his new bride to that northern frontier. Eventually, in 1872, Beames Brook too, was abandoned because of constant strife with the Aborigines. Edkins moved the remaining stock to Mount Cornish, then an outstation of and thirty miles from Bowen Downs. It was situated on treeless, grassy downs country on a branch of the Thomson River. Mount Cornish then became a separate station and ran cattle while Bowen Downs specialised in sheep. Edkins was appointed the manager of this property for the Scottish Australian Company.[20]

Before a manager was appointed to Beames Brook, Nat was busier than ever trying to supervise the two far flung properties. The new station had to be supplied with stock, plant and provisions from Bowen Downs, where cattle-spearing was getting increasingly serious. To add to his busy schedule he had to prepare to take charge of 35,000 sheep en route from Fort Cooper to the Downs.

The sheep were to be delivered in two separate lots. Nat took charge of the first mob at Fort Cooper and set out through drought-stricken country along the track he had pioneered in 1862. This was a nightmare journey. Progress was affected by drought, feed was scarce and the waters further apart than usual. Dingoes waited for any weak or straying animal. Near the Belyando River tragedy struck the sheep when driven by hunger, they ate a poison weed which killed them in their hundreds. Buchanan delivered the first depleted herd and returned immediately to Fort Cooper for the second draft. A large area of poison weed was chopped out to avoid the catastrophe of the first trip, but still losses were quite heavy. From the total of 35,000 sheep that set out, a massive 10,000 died. The final mob was delivered in January 1865, and by some quirk of fate their arrival coincided with a slump in woolprices.[21]

It is uncertain when Nat escorted his wife and baby back to Bowen Downs, but they remained on the station for two years thereafter so it must have taken place in 1864. Kate did tell of another close call in the constant war of nerves waged by the Aborigines against the white settlers, and it seems probable that the incident happened on the return trip to the station.

To prevent surprise attacks it was customary for Nat to eat the evening meal and move on after dark for a mile or so before setting up camp. On this occasion the weather was mild and there was no sign of trouble so the tent was set up. Sometime during the night Nat had a premonition that all was not well, so he woke Kate and moved on for a mile or two and camped under the stars, leaving the tent at the earlier site. Rising before dawn, he returned to the previous camp to collect the gear left behind only to discover the tent riddled with spears. Nat never again relaxed his rule to move on after dark.

Since the Bowen Downs homestead had been moved to Cornish Creek, probably sometime in 1863, the establishment had grown rapidly and there were more than just a few bark huts on the riverbank now. When Rowley Edkins took his wife up to Beames Brook they stopped at Bowen Downs and Edwina Edkins recalls in her memoirs, 'staying in a one room dwelling built for Mrs. Buchanan'. The human population of the station had grown along with the expanding flocks of sheep. Initially it was expedient to stock with cattle because they travelled better than sheep. However, it was always planned that Bowen Downs should be primarily a sheep-grazing property. By 1865 independent bullock drays made twice yearly visits to the downs bringing supplies and news of the outside world from far away Port Denison. Towards the end of 1865, Cornish appointed John Ranken to manage the cattle and William Steele as sheep overseer, and this lightened the load for Nat. In 1866 the price of wool dropped and cattle became unsaleable and the year was also marked by a financial depression and failure of the London Bank which left the Queensland Government almost penniless. This was definitely not an auspicious climate in which to renegotiate the five year agreement between the partners, which was due in October.

Morehead, although acknowledging that Nat was the best man for the job of establishing the property, found it irritating that his manager didn't make frequent written reports. It was the ideal time for Morehead to dismiss Nat and appoint a new manager with whom he could work more closely. The resident manager had served his purpose establishing and organising the stocking of both Bowen Downs and Beames Brook and this special expertise of his was no longer required. Morehead appointed his son Boyd Dunlop Morehead who was later, in 1888, to became Premier of Queensland. The debt of 14,000 pounds outstanding against Nat's one-eighth share in the company was taken over by the Scottish Australian Company, Landsborough and Cornish.[22]

Cornish was dispatched to visit the Company's holdings in the North. This was his last visit to the Gulf Country. The unfortunate man contracted gulf fever and was escorted back to Sydney by Rowley Edkins. Despite the dedicated nursing of his wife Cornish died shortly after his return. Mrs. Cornish became infected with the same disease and her untimely death left eight children orphaned.[23]

It was not long before Landsborough too ran into financial difficulties and was forced to relinquish his interest in Bowen Downs. This left the Scottish Australian Company as the only survivor. Landsborough married in 1862 and had two daughters before his wife died of tuberculosis. He lost Glenprairie and became Police Magistrate and Commissioner for Lands in Carpentaria and was stationed in Burketown. This lawless frontier was policed by W. D'Arcy Uhr with whom the Magistrate was frequently at loggerheads. Landsborough was dismissed from the post under a cloud when he misjudged a legal situation. He returned south and remarried in 1873 and had three sons by his second marriage. This grand explorer and gentleman died in 1886 and is buried at Caloundra.

Nat had spent six years establishing Bowen Downs and he left a thriving small community of some sixty people, including three women, where once there was nothing but rolling downs. With his sights on the challenges of the future rather than the disappointments of the past, he walked away empty-handed but not defeated.

Nat's name is still alive and well in the Longreach area. Bowen Downs has survived many ups and downs over the years but is still flourishing and is a lasting reminder of its founders' foresight and faith in the country they worked so hard to secure.

References

1. Angela I. Moffat, *The Longreach Story, A History of Longreach and Shire*.
2. *The Journal of Landsborough's Expedition from Carpentaria in Search of Burke and Wills*, Facsimile edition. Melbourne, 1862, and letters from S.E. Pearson to Gordon Buchanan dated 17/1/34, 13/2/34.
3. *Australian Dictionary of Biography*, Vol. 5, 1966, Melbourne University Press - William Landsborough.
4. Gwen Trundle, (nd). *W. Landsborough, Explorer*, unpub. m.s., Longreach Qld.
5. Letter from R. Morehead to W. Landsborough 19/12/1866, Landsborough collection, John Oxley Library, Brisbane.
6. W. Landsborough's first diary of the 1860 journey in which this part of the trip would have been recorded is missing. Landsborough's biographer, Gwen Trundle in her unpublished ms. states that Lake Buchanan was discovered and named by Landsborough and Buchanan. D.S. MacMillan in his history of Bowen Downs quotes from a letter written by E. Cornish, Sydney to Charles Grainger, London on 20th May 1862 stating that the lake was discovered and named by Buchanan when making the first road to Bowen Downs.
7. *Rockhampton Bulletin* 25/10/1862.
8. *Sydney Stock and Station Journal*, 25/11/21.
9. E. Palmer, 1903. *Early Days in North Queensland*. Sydney.
10. Letters from S. E. Pearson to Gordon Buchanan dated 26/3/34, Buchanan Collection and E. Palmer, Early Days in North Queensland, Syd 1903.
11. E. Palmer, op. cit.
12. D.S. MacMillan, 1963. *Bowen Downs 1863-1963*. Sydney.
13. Letter from E. Cornish to W. Landsborough 21/1/1864. Landsborough Collection. J. Oxley Library, Brisbane.
14. *Queensland Government Gazette 10/12/1859* - 29/12/1871, Vol.1 No.1 J. Oxley Library, Brisbane.
15. Extract of Marriage Certificate of Nathaniel Buchanan and Catherine Gordon.
16. *The Maryborough Chronicle*, 4/1/1864, 7/1/1864.
17. Extract of Birth Certificate of Gordon Buchanan.
18. Letter from E. Cornish to W. Landsborough, 21/1/1864. op. cit.
19. *The Queenslander*, 17/2/06.
20. Edwina M. Edkins, (nd). *Reminiscences of Edwina M. Edkins.* Unpublished ms. held at Bimbah, Longreach, Qld.
21. D.S. Macmillan, *Bowen Downs 1863-1963*, op.cit. 1963, Sydney
22. Letter from Robert Morehead to W. Landsborough, 19/12/1866. op.cit.
23. Gulf Fever was a form of Dengue, a disease of tropical and sub-tropical regions caused by an arbovirus transmitted to humans by the mosquito.

6. Manager's House Bowen Downs, 1875
John Oxley Library

NAT BUCHANAN

4. Nat Buchanan
Probably taken on the occasion of his marriage in 1863
Buchanan Collection

5. Catherine Buchanan
Probably taken on the occasion of her marriage in 1863
Buchanan Collection

4.

Farming and Tin Mining

1867-1873

———⌒∽∝∝∝∽⌒———

There is no record of what work Nat was engaged in immediately following his time on Bowen Downs. It is possible he took up contract droving or a short stint of station management. Gordie, who could not recall anything of this period, said his first recollections were of a journey from Queensland to visit his maternal grandparents who were on Bective Station near Tamworth.

The weary travellers were warmly welcomed by the Gordon family and to mark the occasion John Gordon presented his grandson with a two year old bay gelding:

> I named him Quondong, and whether he was running brumbies on the western foothills of New England or mustering cattle in the Northern Territory, the longest day was never too long for this game little horse. When he died on Flora Valley Station in East Kimberley in 1898, he must have been thirty or more years old.

Nat and his brother Andrew took up a property under Sir John Robertson's Land Act, on the Bellingen River at Deep Creek and this was their destination.[1] Bidding fond farewells to the family on Bective Station the trio set out for Deep Creek. After leaving the New England tableland, the track became so difficult that the buggy was left behind and they proceeded by horseback and packhorse through the thick forest and scrub that clothed the eastern slopes of the Great Dividing Range. There was only a bridle track to follow and for miles it was impossible to diverge because of the density of the bush. The cedar and blackbutt shut out the sun and the track was often barred by fallen logs which Gordie enjoyed jumping Quondong over. Now unrestricted by the confines of the buggy and able to ride his own horse the boy was enjoying the adventure. Kate rode side saddle in the long riding habit of the period and both adults led a packhorse each. The narrow track made travelling in single file necessary so there was a great mix up when the tenacious barbed lawyer vine caught in Kate's skirts and she became hopelessly entangled. Nat, in an attempt to release her from the clutches of the vine, tore her skirt. Unrepentant, he remarked in his usual unflustered way, 'Oh murder, Kate you will have to ride in your petticoat now.' A pair of kookaburras laughed as they watched the confusion. The songs of the coachwhip birds, bellbirds and butcher birds accompanied them and Nat entertained Gordie by identifying the various bird calls that came from the canopy above.

Life was hard but healthy at Deep Creek, and the living conditions were primitive. The little community included Nat, Andrew, Frank and their now elderly father, plus Kate and Gordie and Andrew's wife Sara and their six children. Mr. Buchanan Senior had made his home with Andy following the death of his wife Anne in 1863.

The rich soil needed only a tickle with the hoe to produce vegetables and maize, and there was plenty of wild honey. Goat, fowl, fish, butter, milk and eggs were regularly on the menu but flour, sugar and tea were often missing. Kempsey was the nearest town and was accessed via a bridle track through the shadowy forest. Getting produce to and from the markets where there were no roads and many river crossings was often impossible for long periods, so the families had to be self-sufficient. Using axes and horses, the job of clearing heavily forested land for cultivation and grazing was backbreaking work.

Nat, used to the wide open spaces of central Queensland, found the confinement of the forest oppressive and he longed to escape to more open country where he could see further than his own pasture. This, plus his dwindling finances gave him the excuse to try his hand at tin mining at Watsons Creek, near Bendemeer in New England. Nat was the first of the brothers to leave Deep Creek and Frank eventually left too. In 1870, Mr. Buchanan Senior became ill and had to go to Sydney for an operation from which he never recovered, but Andy remained on the property and raised a family of seven children.

Although Nat was not demonstrative in his nature, his mood on leaving Deep Creek, according to Gordie, '...was of cheerful optimism'. He was happy to be on the road again and escaping the clammy climate and the endless shadowy scrub and forest.

On that first night rain threatened, so they took the precaution of camping in a deserted hut which cattle had long used as a sheltered camping spot. It needed a good clean-out to make it fit for human habitation, but they were glad of the shelter because it rained heavily until morning. The cattle, however, resented being evicted and kept them awake with their determined and repeated attempts to break down the door.

By next afternoon the muddy, wheel-rutted track brought the travellers to the MacLeay crossing where they were faced with a half banker of yellow water flecked with foam and forest debris. The high banks on either side were topped with forest and although Kempsey was not far off there were no houses in sight. Nat spotted a rowing boat moored over the other side and decided to swim across and bring it back. He and Gordie first walked to a point about three hundred yards up the northern shore where Nat stripped off all but his shirt and waded in. For a while his shirt bellied out like a balloon until it got wet, then Gordie could only see his head making diagonally across the stream to where the boat was moored. He had gauged his own and the current's strength just right. After rowing Kate and the camp gear across first he then swam the horses across in pairs on their halters, behind the boat.

While in Kempsey, Nat was approached by a land agent to look at a 'beautiful river farm' he had for sale. When Nat agreed to look at it, his experienced eye was quick to notice and comment that most of the land was on a flat below high water mark. When the agent vehemently denied this observation, Nat pointed out the flood marks about two feet up the trunks of the trees. These, the agent glibly explained, were the marks made by wild pigs rubbing themselves against the trees. Further on Nat noticed that the flood water marks on the trees had risen to five feet, but he said nothing to the agent. After the inspection the 'land shark' was keen to close a deal and asked, 'Now Mr. Buchanan, wouldn't you like to own some of this rich land?'

Nat, with a good humoured twinkle in his eye replied thoughtfully, 'No, but I would like to own some of those pigs!'

With their supplies replenished in Kempsey the family headed west, threading their way through the scrub to reach the foothills of the great range. Then, as now, it was an imposing barrier but in those days the roads were rough and some were only bridle tracks. A bit of the scrub had been partly cleared for cultivation and the north east plateau was scaled here and there by bridle tracks which followed marked tree lines. From some of its summits there was an uninterrupted spectacular view to the Pacific Ocean. In earlier years the area had been inhabited by independent Aborigines, but they had since fallen victim to the harsh reality of European settlement and the way was now uncontested. Slowly the last steep grades of the old mountain were negotiated and near the summit they pitched camp for a day or two to shoot some game and enjoy the bracing air of the tablelands.

Near Manila the family dropped in on Kate's sister Jessie and her husband George Farquharson. After a short but happy family reunion the travellers followed down the crystal clear McDonald River to granite-bound Bendemeer. This pretty, sleepy hollow hasn't changed much over the years. Glover's Inn, once a model of respectability and comfort, has had additions made but the ground floor structure is the original. In Nat's time there was a police station, some government offices and a few private cottages. The river was then, and still is, spanned by a fine white bridge, and the banks are lined with shady waving willows. The discovery of the Watsons Creek tinfields disrupted the tranquillity of the little town for a short period and brought some prosperity to the area.

In the swampy terrain between Longford Station and Watsons Creek the party was forced to slow down and it was here they were startled by several gun shots which came from a thick patch of timber not far ahead. Concerned that it might be bushrangers, Nat was preparing to take action when two men rode out of the bush firing at each other. Nat shouted and one rider galloped off while the other fell wounded from his horse. By the time Nat reached the wounded rider he was sitting up and his horse had moved off to graze nearby. Kate hurried to the fallen man and after examining his wound found a bullet had passed through the top muscles of his shoulder. When Nat had brought her some water, she washed and bandaged the wound, but was unable to extract the man's name or much information about him. He was obviously keen to keep his assailant's and his own identity concealed. All the man would say was that he was lucky to get off so lightly, and after thanking them he rode off in the direction of Bendemeer. Much later Nat heard on the bush telegraph that the row was over a woman.

Watsons Creek was a tiny village with a store and a butchery which was operated by Joe Steele. Initially the family set up home in a tent but soon Nat, with the help of an Irishman named Connaught Jack, erected a bark and slab hut on the bank of a running creek about a mile from the village. The hut had two rooms, a kitchen and a skillion under the one weather-proof roof. The verandah fronted on a deep waterhole and beyond the creek was an open flat. At the back door there was thick forest which shaded the hut from the afternoon sun and was home to wallabies and possums. Ti-tree lined the banks of the creek and Kate put this to good use to make her household brooms.

The bridle track between Watsons Creek and the Giant's Den, where lode tin was being mined, crossed the creek lower down and in sight of the house. From the front verandah Gordie often enjoyed watching parties of Chinese crossing the creek, some on horseback and some on foot, laden with the inevitable pole and baskets. Gordie was fascinated by the newly arrived immigrants because their appearance and habits were so different. Previously he had only seen individual Chinese as cooks, gardeners or shepherds on the stations. These strangers were indifferent riders and rode in single file at a slow walk. Much to Gordie's amusement the lead rider would occasionally kick his horse into a sluggish trot and the other horses would follow suit. There would be a desperate clinging to mane and pommel and frantic, high pitched shouts arising it seemed, from a mixture of both terror and amusement.

Nat employed an English-speaking Chinese as overseer of the Chinese workers on the alluvial tin claim, half a mile away. Part of the Watsons Creek flow was diverted at a point five or six hundred yards higher up the stream and then by gravitation travelled along a deep trench to the head of the workings. There the wash dirt was treated in cradles or sluices of some kind, by the head of crystal clear water. The tin was packed in twelve by eight inch canvas bags, each weighing about 70 pounds.

Nat did quite well for a while but all too soon, either from the increased difficulties of mining it or a fall in prices, the tin failed. Joe Steele and Nat formed a brumby-running syndicate and with the help of his nephew Andy and Gordie, then nearly ten years old, set about catching some of the many brumbies which thrived on the western slopes. The brumbies liked to graze about ten or twelve miles away where they enjoyed the short sweet grasses in the open forests of white and yellow box. Gordie, riding Quondong, had the job of tailing the coachers near a trap yard into which they sometimes forced a few wild horses who were too jammed in by the coachers to escape. There were generally some breakaways which it was useless to pursue. One day Gordie saw a mob of brumbies led by an imposing chestnut stallion, appear over a nearby ridge. When the stallion saw Gordie he pulled up dead, then gave a loud whistling snort and with toss of head and mane, disappeared over the hill with his mob. Eventually he was yarded but fell to the lot of Joe Steele. First pick of the horses was always decided by drawing straws, longest straw was awarded first pick. Because too many others were at it the brumby-running game didn't pay, so once again the Buchanans' moved on.

References

1.   Sir John Robertson was a land reformer and politician. On the 24th October, 1861 assent was given by the New South Wales Parliament for his two land reform bills. '...unsurveyed land could now be selected and bought freehold in 320 acre lots at one pound per acre, on a deposit of 5/- per acre, the balance to be paid in 3 years, an interest free loan of three quarters of the price; and that wealthy people would find it hard to speculate because bona fide residence was stipulated.' *Australian Dictionary of Biography*, Vol. 6, 1966, Melbourne University Press - Sir John Robertson.

5.

Life on Craven Station

1874-1875

~~∞~~

Nat intended to find land for running sheep and cattle under Sir John Robertson's Land Act, but when he couldn't find any to suit him, accepted instead the management of Craven Station in Queensland, for the Bank of New South Wales.

The brumby-running horses and outfit were sold to Joe Steele and Gordie reluctantly parted with Quondong, who was left in the care of his uncle Hugh, then manager of Millie South. From Tamworth the family took the coach to Singleton then the train to Newcastle, where they boarded a steamer for Rockhampton. The coach trip from Rockhampton to Clermont was a wearisome experience. Nat and Gordie resented being cooped up and Kate stoically bore the discomfort as the vehicle lumbered and jolted along on its leather springs all day and into the night. Facilities provided at the midday and night stage houses along the route were primitive. Often there wasn't enough water to rid hands and faces of the dust which lay as thickly over the passengers as it did over Her Majesty's mail. However, the food though plain, was plentiful.

At Clermont they were relieved to be met by the outgoing manager of Craven, Mr. Green, who carried them the remaining twenty or thirty miles in the relative comfort of the station buggy. Part of the way lay through forest and scrub and the track twisted and turned to avoid areas of 'dead finish' and brigalow. The red soil, churned into dust by heavy wheel traffic, clung to everything, including the green foliage beside the track.

At the station they were cordially received and Kate was pleasantly surprised to find her new home was a comfortable and commodious cottage. There were four rooms surrounded by cool, wide verandahs and a garden of roses, oleanders and Isabella grapes. Kate was delighted. A Chinese gardener grew vegetables down at the creek near a big waterhole and further on there was a camp of Aborigines. Joe Hobbs was the headstockman and a man and his wife and daughter comprised the home staff. Wild lime trees grew in abundance and the fruit was frequently gathered by Kate to make delicious drinks in the hot summer. At the woolshed and elsewhere iron tanks collected water, but in the dry times a fork of a tree was used as a sled to draw water from the creek about four hundred yards away. Beside the woolshed and men's quarters there was a store and smithy, an office and a dairy, and a shed for meadow hay. The property was mainly a sheep run with a few cattle and was unfenced except for a small horse paddock. There was a mixture of good and bad country, typical of land on the upper Belyando tributaries, and although there were no mountains and few hills, sheep were continually being lost and shepherding was hazardous.

The Aborigines of the Belyando and Dawson were many and bold, and had a history of attacking whites. They had the advantage of secluded hiding places in the thick scrub and numerous creeks. In response to ongoing depredations a troop of mounted black police was stationed nearby. The troop consisted of from four to ten recruited Aborigines under the command of a white police officer, Inspector Nolan. His headquarters were a few miles down Craven Creek which was about the centre of his district and the most disturbed part of it. Nolan gave the raiders constant and often well deserved attention. His troopers rarely, if ever, brought in a prisoner unless the Inspector was with them, and probably not always then.

Most of the shepherds were Chinese and because of their isolation bore the brunt of Aboriginal attacks. When a Chinese shepherd was speared, the troopers would be sent out to get the culprits by any means. After a few days or weeks the troopers would return and provide only a vague account of their activities, but the Inspector would find all the evidence he required by the notches in their rifle butts.

'Here! Peter, Billy, Spider! What for you shootem that one blackfellow belonga Alpha Station mob?'

The typical answer came back, 'We bin tellum boss,' stan three fella time longa Queen,' but him bloody fool, he no understan' white language.' Unfortunately the Aboriginal tribes understood too well the language of the rifle.

The terror of lonely shepherds was Belyando Billy, alias Long Billy, alias Birrianda. Their heads were never safe from his attentions and although Birrianda had been wounded once in a 'dispersal', there was no notch on any carbine for him.

Birrianda had formerly been a police trooper, so this experience plus his cunning gave him an advantage. He rarely slept in the main camp, but some distance away with his women who remained vigilant day and night to give him early warning. He wore mocassins of grass or emu feathers to make tracking difficult, and would slip away unnoticed into his favourite patches of scrub, so thick according to Joe Hobbs that, 'A dog couldn't bark in it'.

Birrianda was lucky, too, because he had an ally in a friendly but eccentric white shepherd who gave him food and shelter. The shepherd was about sixty years old and according to rumour he was a relative of a high Anglican dignitary in England. Although probably suffering from a psychiatric disorder, he was nevertheless an excellent shepherd, the best on Craven. His flock was always out by daybreak and in by dark, and he judiciously varied the grazing grounds and didn't harass the sheep by needless dogging. The shepherds didn't have horses and he was adept at turning the flock himself, on foot. One of his eccentricities was to discard his clothes for days and daub himself with mud from the waterholes, but whatever his dress he always carried a Bible. His madness together with his sympathetic treatment of Birrianda saved him from harm at the hands of the Aborigines. Usually he had an entourage of them with whom he shared his fortnightly rations of food and tobacco. Eventually his increasing madness became too great a responsibility for the station and he was institutionalised. With his benefactor gone, Birrianda took some of his little used clothes and left the district, and ended up working on Bowen Downs. Although his identity was suspected, no attempt was ever made by police or civilians to 'put him away.'

In a short time Craven management realised that the country was not suitable or safe for sheep, let alone shepherds, and along with Banchory and Beaufort Stations they changed over to running cattle. Around mid-1875, Craven again changed owners and made possible the fulfillment of Nat's desire to strike out again on his own. He was not unduly restricted in his management of Craven, and the open and wider range of activity partly satisfied his inherent restlessness, but he could no longer resist the challenge of the unknown. The opportunity came when Benjamin Crossthwaite of Melbourne asked Nat to pilot cattle to newly leased country on the head of the Herbert River (Georgina) in the Northern Territory of South Australia.[1]

Nat expected to be away from Kate and Gordie for twelve months during which time they were to live in a rented cottage in Copperfield. After goodbyes were said to their friends on Banchory and Beaufort stations they set out in the buggy for the mining village. One can only guess how Kate must have felt. She was leaving a comfortable, secure home and friendly neighbours to live in a strange town where she would have the responsibility of bringing up her son in his father's absence. Copperfield was a long way from New England, so she could not depend on her family for support. As a pioneer in her own right she understood Nat's need to keep reaching further out, and her own bush experience told her that he could not do this effectively or safely, encumbered with a wife and child.

As the horse trotted along the road to Copperfield at a smart pace their equilibrium was suddenly lost as the off-hind wheel flew off, ejecting Nat onto the road. Kate reacted quickly, grabbing the reins in time to steady the pair, while the dragging axle acted as a brake. The problem was caused by a too tightly washered wheel. When it broke away it took all the screw thread with it, leaving enough axle arm to support the wheel but not enough for security. Nat's skill at inventing makeshifts came to the fore. A sapling was cut and the butt end strapped to the fore carriage so that the middle of it lay against the outside rim of the wheel, to prevent it coming off. The trick worked well, except two or three saplings wore through on the journey and had to be replaced. Saplings were ideal for the job and in plentiful supply along the track; heavier or dry poles would have damaged the wheel.

Copperfield was a small town near Clermont which grew up as a result of the discovery of copper there around 1861. Copper mining was at its height of production by 1872, but prices crashed in 1873, so when Kate and Gordie arrived, people were leaving the town and there were cottages available.[2] Nat saw his family safely settled and provided funds to draw upon while he was away. On September 3rd 1875, he rode west with Crossthwaite to look out and settle new country on the Tablelands west of the Darr and Diamantina watersheds.[3] From this time forward Nat was destined to travel the uncharted tracks across the north of Australia, without a home to call his own.

References.

1. *The Sydney Morning Herald*, 1/5/1876.
2. Kerry Killen, 1984. *Drovers, Diggers and Draglines*. Pub. Pacific Coal Pty. Ltd.
3. *The Sydney Morning Herald*, 1/5/1876.

7. Tennis party at Rocklands Station
John Oxley Library collection, Northern Territory Library, PH 171/157

6.

The Western Queensland Frontier

1875-1877

On the 3rd. of September, 1875, Messrs N. Buchanan and Crossthaite
[sic] left Copperfield to look out and settle country on the fine
tablelands to the west of the Darr and Diamantina watersheds.

Sydney Morning Herald, 1/5/1876.

Mount Cornish station was their first stop. Here Ben Crossthwaite had
arranged to buy cattle to stock the country he had leased on Lake Mary. Rowley
Edkins, who was well known to Nat, was now the manager of Mount Cornish,
which had originated as an outstation of Bowen Downs. Crossthwaite's intention
was to continue with the cattle into the little known country west of the Darr and
Diamantina watershed, and then on to Rocklands which lay close to the present
town of Camooweal. However, a shortage of surface water for the cattle forced a
change of plans and the cattle were left on land recently taken up by D.S.
Wallace.[1] The property was Elderslie and the head station, or 'Wallace's Camp' as
it was then called, was situated on Pelican Waterhole about 480 miles from
Copperfield.[2]

The small party crossed from the head of the Diamantina to the Cloncurry and then because water was still scarce, rode north up this river and crossed to the Leichhardt. From here they made a second attempt to go due west but their unshod horses got footsore due to the rough country so they were compelled to backtrack. Once more a northerly course was adopted, this time travelling up the Leichhardt River until it was possible to head west to the Gregory and the O'Shaunessy Rivers.[3]

Arriving finally at Rocklands they made their depot on Lake Mary. Nat and Ben took a few days rest in this idyllic spot to recover from the rigours of their journey. The lake was beautiful, about two hundred yards across and three quarters of a mile long. It had a high western bank on one side and vast undulating downs on the other. Rocklands was 200 miles south west of Burketown, which at that time was a ghost town because Gulf Fever had decimated the population.

Rocklands was first settled by John Sutherland in November 1864, when he arrived with 8,000 sheep bought near Rockhampton. It was Sutherland's intention to find new land in the Gulf country to stock but all the good land was taken up and the rest was unsuitable for sheep because of the coarse grass and grass seeds. Landsborough's account of the splendid grazing lands of the Barkly Tablelands inspired Sutherland to continue west until he came to Lake Mary. Here he set up a permanent camp. The lack of water, distance from suitable markets and a crash of stock and wool prices caused him to abandon the new station after a few years.[4] Rocklands spans the Queensland, Territory border but as this settlement took place well before the demarcation of the border, the settlers were unsure in which colony they resided.[5]

William Landsborough was out on the Barkly in December 1861 and Nat soon recognised blazes on some trees, marked by his old friend. While in search of Burke and Wills, Landsborough had attempted to cross the Barkly but was driven back by lack of water.[6] Nat and Ben Crossthwaite made exploratory trips to the west of Lake Mary as far as the Ranken River. This river was named in honour of Nat's friend and contemporary, John Logan Campbell Ranken, who first settled on Avon Downs in 1866.[7] W.O. Hodgkinson, in his report on the North West Expedition, mentioned meeting Nat near Lake Mary and recorded that he had explored and named the Ranken.

Hodgkinson led a Queensland Government expedition in 1875/76 to assess the amount of pastoral land existing between the Diamantina and the border, and its geological capacity to produce minerals. His second in charge was a mining surveyor and minerologist named E. A. Kayzer. When their mission was nearing its end in October 1876, Nat came to the assistance of the weary party and their knocked up horses. Heading for Normanton with an empty dray to get supplies for Rocklands, he was able to transport the expedition's packsaddles and mining survey instruments.[8]

The sort of country that impressed Nat was well grassed, tree studded, downs country. What he saw of the Barkly Tablelands on his exploratory trips suited him well. He was never attracted to the heavily timbered slopes of the higher rainfall areas. Nat was keen to secure land on the Herbert River (Georgina) in the Northern Territory of South Australia because it could be leased without immediate stocking. However, when he explored the area he was disappointed to discover that the Herbert (Georgina) only entered the Territory for a few miles before turning back into Queensland. He did, however, apply for some leasehold land in the Territory as early as April 1876, but for reasons best known to himself, never went ahead with it.[9]

While still out west, Nat and a trusted Aborigine named Jimmy made a 300 mile trip to examine the source of the Gregory and Nicholson Rivers. When searching out and stocking new country, he made many unrecorded scouting expeditions. One incredible incident emerged from this trip and is probably the reason that it was recorded.

It was a calm, clear day and Nat was sitting on a river bank, fishing in the mirror-like water. To his dismay, he saw the reflection of an Aborigine standing behind him with a spear poised ready to throw. While he was frantically trying to think of some avenue of escape, 'lady luck' intervened and a large fish tugged at his line. Continuing as if unaware of the threat immediately behind him, he landed the fish and then turned around quite casually and with a smile, offered it to his would-be assassin. The Aborigine was completely disarmed by this spontaneous gesture of friendship.

Following good rains Nat and Crossthwaite left Rocklands on the 15th February 1876 to collect the cattle waiting on Pelican Waterhole and move them to Rocklands. Between leaving the cattle and his return Nat had travelled 1,500 miles in three months and three weeks.[10] Today this seems an incredible feat when we bear in mind that the country was largely uncharted, there were no roads or towns, and travel was on horseback through rough country where whites were unwelcome.

At the same time as Nat was associated with Crossthwaite he went into partnership with a man named Kirwin. Their property, named Herbert Downs, was situated on the Herbert River between Glenormiston and the present day Boulia.[11] Not all the cattle he collected were for Rocklands, some were probably to stock Herbert Downs:

> The partners acquired cattle to stock the station but a series of calamities, culminating in the death of Kirwin within the first year of occupation, caused Nat to abandon it.

The disaster that followed Kirwin's death is vaguely documented and Gordie's two references to it differ slightly. What appears to have happened is as follows. Kirwin, Nat, two white stockmen, Jimmy and two other Aborigines made up the complement of men on the fledgling station. Supplies ran low and there was no salt, so to save waste, calves instead of adult cattle were slaughtered for meat. This diet caused dysentery among the men, so Kirwin took the dray and set out to get supplies from Normanton while Nat and Jimmy borrowed a few supplies from a distant neighbour to keep them going. When Kirwin was overdue, Nat and Jimmy went to look for him and found him dead. It seems Kirwin was delayed by floods on the return journey and died of dysentery. After burying the unfortunate man they drove the loaded dray back to the station. Upon their arrival they found Jack Parker, one of the stockmen, barricaded inside the hut and in fear of his life.

The two stockmen had been out on the run and about midday Parker was busy among the packsaddles while his mate had a quiet doze under the shade of a coolibah tree. From the cover provided by a wide patch of lignum, a rain of spears hurtled towards the unsuspecting men. One found its mark in the sleeping stockman. Parker partly parried a spear, grabbed his revolver and shot one of the Aborigines dead. Several other heads appeared but vanished as he fired at the nearest. When all were out of sight, he turned and called to his mate. He stared aghast at the bloody spectacle which met his gaze. The prostrate body of his offsider was transfixed by a stone-headed spear. Now alone, Parker took his rifle and searched the lignum for the attackers but they had melted into the bush. Returning, he covered his mate's body with a blanket, attended his own painful wound, and then limped off with rifle and bridle to fetch the horses. Upon his return he discovered that the Aborigines had visited the camp in his absence, collected their dead brother and stolen all the rations. With tomahawk, a prospector's tin dish and an improvised yam stick, he arduously scooped out a shallow grave, wrapped the dead man in his blanket and buried him, and got away from the scene by sundown. Once back inside the hut he had to set up barricades against an attack by a large number of Aborigines. Parker was forced from his shelter under cover of night to make the perilous trip to the lignum-lined billabong for water. The horses were spared, but the cattle were speared and scattered far and wide by the Aborigines.

Parker explained to Nat that soon after his departure the two Aborigines had deserted and although he didn't think they were the ringleaders he blamed them for instigating the attack. In later years the two offenders maintained that they had been provoked by the white stockmen.

> *Clashes between the Aboriginal owners of the land and the whites were inevitable, yet there were some, like Nat, who respected their native bushcraft and often relied upon their goodwill. He understood enough of their nature to deal with them in such a way that he was able to penetrate their land peacefully. The majority of pastoral pioneers antagonised the tribes with their ruthless severity, making them suspicious, cunning, and ready to take revenge.*

Another of Nat's attempts to establish himself and provide a home for his family had failed. When he walked off Herbert Downs, he parted company with his loyal servant Jimmy, whom he returned to his tribe near Normanton.

Temporarily frustrated, he made the long trip home to Copperfield via Mt Cornish, where his friend Rowley Edkins lent him a change of horses for the remaining 150 miles. Even though he had been out bush for twelve months he only took a short break because he was keen to continue the search for better country across the border.

References

1. *The Sydney Morning Herald*, 1/5/1876.
2. *Winton Queensland - Originally Pelican Waterhole, 100 Years of Settlement 1875-1975*, Shire of Winton.
3. *The Sydney Morning Herald*, 1/5/1876.
4. George Sutherland, 1913, *Pioneering Days: Thrilling Incidents Across the Wilds of Queensland with Sheep to the Northern Territory in the Early Sixties*, W.H. Wendt & Co. Ltd. Printers, Brisbane.
5. The surveying and marking of the border between the Northern Territory and Queensland commenced in early 1884 and ended in Burketown in October 1886. Prior to this survey property owners adjacent to the border were uncertain as to the colony in which they resided. The Demarkation Party consisted of Mr Poeppel the leader and L.A. Wells as second in command, Mr. Woodward the Assistant Government Geologist, six assistants, and two Afghan cameleers.
6. *Australian Dictionary of Biography Vol. 5.* 1966. Melbourne University Press. - Ref. W. Landsborough.
7. Carmel Wagstaff, Avon Downs N.T.
8. *The Queenslander*, 6/7/1878
9. N.T.R.S. F790 A1517, S.A. Records Office, Adelaide S.A.
10. *The Sydney Morning Herald*, 1/5/1876.
11 *The Sydney Morning Herald*, 30/12/1921.

7.

First across the Barkly

1877

No white man succeeded in crossing the Barkly Tablelands from Queensland to the Overland Telegraph line until 1877. Nat had already explored from Rocklands to the Ranken River and was familiar with the Barkly country so he was confident that he could succeed where others had failed. The challenge was impossible for a man like Nat to resist and the prospect of finding good grazing land gave him an even greater incentive. Knowing that other attempts using a more southerly route had been unsuccessful, he decided to follow a north west course.

W. Landsborough named the Tablelands after the Governor of Victoria. When searching for Burke and Wills he wrote in his journal,'..a plain with a tableland of the richest soil and with grasses of the most fattening nature. This tableland I have named Barkly Plains, after his excellency Sir Henry Barkly'.[1] Landsborough failed in his attempt to cross because of lack of water.

A number of parties were equipped to examine the Barkly country. All ran into difficulty. Attracted by the excellent terms offered to pastoral lessees in the Northern Territory the Prout brothers went looking for land and never returned. W. Carr-Boyd, who had been a member of Hodgkinson's North West Expedition in 1875-6, searched for and discovered the remains of one of the brothers and a fragment of a diary. This recorded them as going deep into South Australian territory and it is assumed they died of thirst on their return journey.[2]

A surveyor and late Commissioner for Crown Lands for the Mitchell and Gregory North districts, Frank Scarr, was engaged by a group of Queensland squatters to examine and report on the country they had applied for in the Northern Territory. His journey was expected to take six or seven months, and the crossing was to take place between the 21st and 26th parallel. Lack of water forced his party back and they eventually made their crossing in 1878 on the 19th parallel, far north of their planned route.[3]

About the same time Ernest Favenc, with surveyor S.G. Briggs and G.R. Hedley, set out on The Queenslander Transcontinental Expedition. This was sponsored by *The Queenslander* newspaper and its mission was to find a practicable route from Blackall to Port Darwin for the purpose of the construction of a railway line. Although he did have to deviate from his course, Favenc accomplished his mission and was able to return a favourable report regarding the suitability of the country for a railway.[4]

In 1877, before any of these crossings were attempted, Nat Buchanan, Samuel Burns Croker, and an associate of Crossthwaite's in Rocklands, named W. Tetley, made a swift and successful first crossing of the Barkly Tablelands. Their party was small, travelled light and unlike subsequent expeditions, had no wealthy backers. There is no mention in the available record that the party was accompanied by an Aborigine, which was unusual for an exploring party such as this.[5]

Nat first met Sam Croker on Rocklands Station when he was a member of Hodgkinson's Expedition. Sam was 25 years old, fair, lean and of medium height. Neither a great horseman nor stockman, he was nevertheless a skilled bushman and adept at living off the land. He proved this in 1882 when he ran out of rations and managed to survive for two weeks on snakes and lillies.[6] He was keen on leather work and plaiting which earned him the nickname 'Green Hide' Sam. Sometimes Sam wrote articles for *The Queenslander* under this nom de plume. Nat was a good judge of men and he undoubtedly chose Sam for this expedition because of his skills as a bushman. Croker was just the sort of bloke to have along when facing the unknown, an explorer at heart, he was competent, absolutely reliable and not afraid to face danger. Sam, who was a Queenslander, later earned a reputation for being hard on the Aborigines. Perhaps he had been influenced by the summary justice handed out to Aborigines by the native police in his home State. When with Nat, it seems that Sam deferred to the older man's more peaceful methods. The trio set out from Rocklands Station on the 10th October 1877. The country they travelled was flat, open grassland which, owing to an especially good season that year, waved in the breeze like a vast wheatfield. They halted at one solitary tree and no other object, tree, bush or hill was in view right around the 360 degrees. In the first ninety miles of travelling they crossed four creeks which fed into the Herbert (Georgina).

> *We then spelled for one day and on the next, after a stage of 29 miles on a north-westerly course, made a creek which we named after our leader, Mr. Buchanan.*
>
> Green Hide Sam,
> The Queenslander, 29/6/1878.

Following Buchanan Creek west for a further thirty five miles the party discovered two lakes and from this point the waters from Buchanan Creek, Playford River and others flooded out on a central depression into lagoons and lakes which teemed with wildlife: They had great trouble getting out of a thick lignum scrub in the area and this halted progress for a time. Croker Lagoons and Tetley's Creek were named in this area but unfortunately don't appear to have been translated onto maps at the time or of the present.

The country in the vicinity of this flooded region was populated by Myall Aborigines. They were tall, well built fellows and unlike those encountered by Landsborough, not particularly hostile. A local tribal custom was to knock out the two front teeth as a sign of reaching manhood and it was also their habit to chew 'pituri', or native tobacco. They combined this with ashes to form a ball which was held in the mouth between the teeth and lower lip or stored behind the ear when not in use. Some were very shy but others approached and were allowed to examine both horses and men closely. At first they thought that horse and man were one and were astonished when the riders dismounted and proved to be similar to themselves. Elsewhere, other parties were warlike. When conciliatory overtures failed and boomerangs and spears whistled dangerously close, warning shots were fired to discourage further unfriendly advances.

Further on, the party found water at Attack Creek, but the next thirty five miles through country covered with a large variety of native shrubs was waterless. That night, after a long day's ride they had a dry camp but noticed some Aboriginal campfires in the distance. Next day contact was made, and the Aborigines obligingly took them to one of their wells where they were permitted to water themselves and the horses.

The three men were jubilant when five miles further on they intersected the Overland Telegraph line near Powell Creek. As no white men had crossed from the east, one can only imagine the surprise and disbelief on the faces of the telegraph station workers. This first successful crossing of the Barkly Tablelands was made approximately along the 19th parallel of latitude.

Nat was impressed with the Barkly country and determined to secure a lease, so he wired his request, which cost eight shillings for twelve telegraphed words. The reply was disappointing. City men had applied for the best country prior to the exploring party setting out. It seems that Nat had no up-to-date knowledge of this 'rush for grass' by speculators. Needless to say, he had achieved what he had set out to do - cross the Barkly - and this success possibly tempered his frustration at missing out on the best grazing land.

A poorly watered, but otherwise reasonable block, was eventually secured by Nat in partnership with someone else. Buchanan Downs, as the property was called, was near the Overland Telegraph Line.[7] It was never developed or stocked by Nat who sold it to Walter Douglas in about 1881 to raise money to stock another property. The present Banka Banka Station incorporates part of the original Buchanan Downs land.[8]

> *The country he explored, but of which no portion could he secure, is now occupied by the well-known cattle-stations, Brunette, Alexandria, Alroy, and Avon Downs, which now carry some of the largest herds in Australia.*

Land on the Barkly was quickly settled and stocked. Of course, some properties such as Rocklands, which is partly in the Territory and partly in Queensland and Avon Downs were settled in the mid-sixties and stocked with sheep. They were abandoned late in that same decade due to financial crises caused by distance from markets, falling wool prices and high freight costs.

Rocklands appears to have changed hands a few times and in 1884 the owners were McCullock, Scarr and Co., and they ran about 7,000 head of cattle. What was called Rocklands Township Reserve became the border town of Camooweal in 1884 when a large store was established and by the next year there were two stores, two public houses, a blacksmith, butcher and approximately fifty residents. There was a fortnightly mail service but no police, magistrate or customs official which made it a haven for the lawless. Sly grog sellers proliferated, horse thieves thrived and a huge quantity of goods crossed the border without the benefit of duty. Large numbers of overlanders passed through the area heading for the Halls Creek gold fields, many thumbing their noses at the law and helping themselves to horses and faking their brands.[9]

Many small properties were taken up by battlers and settled on the Barkly during the 1880's, and along with the bigger concerns, suffered the tyranny of distance from markets, poor water supplies, high freight prices and enormous establishment costs, for which they saw no return. They petitioned without success for a port at Borroloola as an outlet for stock, improved communications, rental reductions and the establishment of artesian bores. Most stock had to travel huge distances to southern markets and it took a minimum of six weeks for mail to reach southern states via Burketown and another six weeks to receive a reply. The pastoralists requested that the Government extend the telegraph service from the Overland Telegraph line across the Barkly to Camooweal, but it didn't happen. Most stations used their own teams to cart supplies from Burketown because the cost of cartage was thirty five pounds per ton.[10] In dry years landholders were sometimes forced to agist stock in Queensland in order to save them, which was frustrating when they were aware that artesian water could be obtained. At Rocklands, artesian water had been discovered around 1885, at a depth of 150 feet and the supply was described as unlimited.[11]

Nat Buchanan's crossing of the Barkly paved the way for this new wave of settlement and opened up a shorter route for stock travelling into the Territory from Queensland and southern states. Until the artesian bores were eventually established along the route in the early 1920's the use of the Barkly stockroute was limited to good seasons. In dry years drovers bound for the east from the north, west and centre of the Territory and those bringing stock into those areas of the Territory via Queensland chose the better watered, but longer Gulf Track.

The Barkly stockroute started at Newcastle Waters crossed Anthony's Lagoon, Brunette and Alexandria before entering Queensland either via Camooweal or south via Lake Nash. Before the sinking of bores the track between Newcastle Waters and Anthony's Lagoon only provided one permanent waterhole and between Anthony's Lagoon and Lake Nash in a dry year there were several stretches of track unwatered for forty miles and more. Cattle destined for Maree in South Australia crossed the border and followed the Herbert (Georgina) through western Queensland to Birdsville and then entered South Australia at Hergott Springs.[12]

There were many interesting and colourful personalities who blazed the tracks 'Out where the dead men lie.' The Barkly Tablelands was not opened up to settlement by the big names in exploring of the day but by stockmen, drovers and pastoralists who were willing to risk their lives and their wealth to settle this promising but unforgiving country. Explorers Ernest Favenc and David Lindsay did some surveying, but it was mainly private individuals who paved the way to settlement. Sadly, today the contribution made by these ordinary men is largely unacknowledged and forgotten and many colourful stories have been lost. However a few anecdotes remain to give us a good indication of how tough life was on the Barkly in those early years of settlement.

Harry Redford, the hero of Rolfe Boldrewood's classic, *Robbery Under Arms*, was a larger than life Tablelands character. He was a bearded man standing six feet tall and weighing around eighteen stone. This experienced bushman, stockman, drover, cattle duffer and horse thief, had frequent brushes with the law in his early years. His greatest exploit was duffing one thousand head of cattle from the Bowen Downs outstation, Mt. Cornish, and overlanding them to South Australia for sale. A reward of 300 pounds was put up by the Bowen Downs Company for his capture and he was eventually arrested and tried at Roma. His brilliant and audacious execution of the crime was so admired by his peers in the jury that they found him not guilty. The judge was disgusted at this miscarriage of justice and he refused to officiate at the Roma Court's criminal proceedings for some months thereafter. Redford continued to engage in nefarious activities and had frequent brushes with the law.[13]

Nat returned across the Barkly to Rocklands following his trail-blazing trip, but there is no record of who rode with him, probably only Tetley because he had an interest in Rocklands. Sam Croker appears to have remained in the Territory because he met Nat next at Katherine in December 1878.

Nat Buchanan's ability to cover long distances on horseback in a remarkably short time was well known, and after reaching Rocklands he took the 700 miles to Copperfield in his stride.

> These long trips, with from three to seven horses, and generally a black boy, but sometimes alone, besides being a severe test of endurance for man and horse, required a good knowledge of the capabilities of his horses and of the best way to sustain their condition. His pace was usually a slow trot or fast jog, sitting loosely in the saddle all the while, easy for the horse but not so easy for the rider. A hack that had a good action trot was much prized. On shorter journeys, and if pressed for time, the travelling hours were lengthened far into the night, and sometimes all night. But in the latter case a short rest was taken every four or five hours. Even twenty minutes for a roll and bite of grass is a wonderful refresher for a jaded nag. For hundreds of miles there was no road or track, and at night the stars or moon were the only guides on the solitary way. The eager hospitality of the few stations en route, all hungry for human intercourse, was always gladly accepted.

As on previous occasions, Nat broke his trip at Mt Cornish Station where he enjoyed the hospitality of the Edkins family and secured a change of horses. Mrs Edkins had a piano and her recitals were very popular with her guests. At Craven Station he joined forces with his brothers-in-law, Hugh and Wattie Gordon, and they rode on to Copperfield together.

References.

1. *Journal of Landsborough's Expedition from Carpentaria, in Search of Burke and Wills*, Melb, FF Bailliere 1862.
2. E. Favenc, *History of Australian Exploration 1788-1888*, Syd. 1888.
3. *The Queenslander*, 18/9/1878, 5/10/1878.
4. E. Favenc, op. cit.
5. *The Queenslander*, 29/6/1878.
6. *Northern Territory Times and Gazette*, 22/7/1882. G. Buchanan, *Packhorse & Waterhole*, 1933.
7. S. A. Archives, NTRS F790 A2722.
8. G. Buchanan, unpub. ms. *Old Bluey*
9. *Northern Territory Government Resident's Report* 30/3/1885 and 30/6/1886
10. *Northern Territory Government Resident's Report* 1884 - 1886.
11. *Northern Territory Government Resident's Report* 31/3/1885.
12. Duncan Ross, 1967, *The Northern Territory Pastoral Industry 1863-1910*, Melbourne University Press.
13. *Northern Territory Dictionary of Biography*, Vol 1, to 1945 edited by Carment, Maynard and Powell, NTU Press, 1990 - Entry Harry Redford.

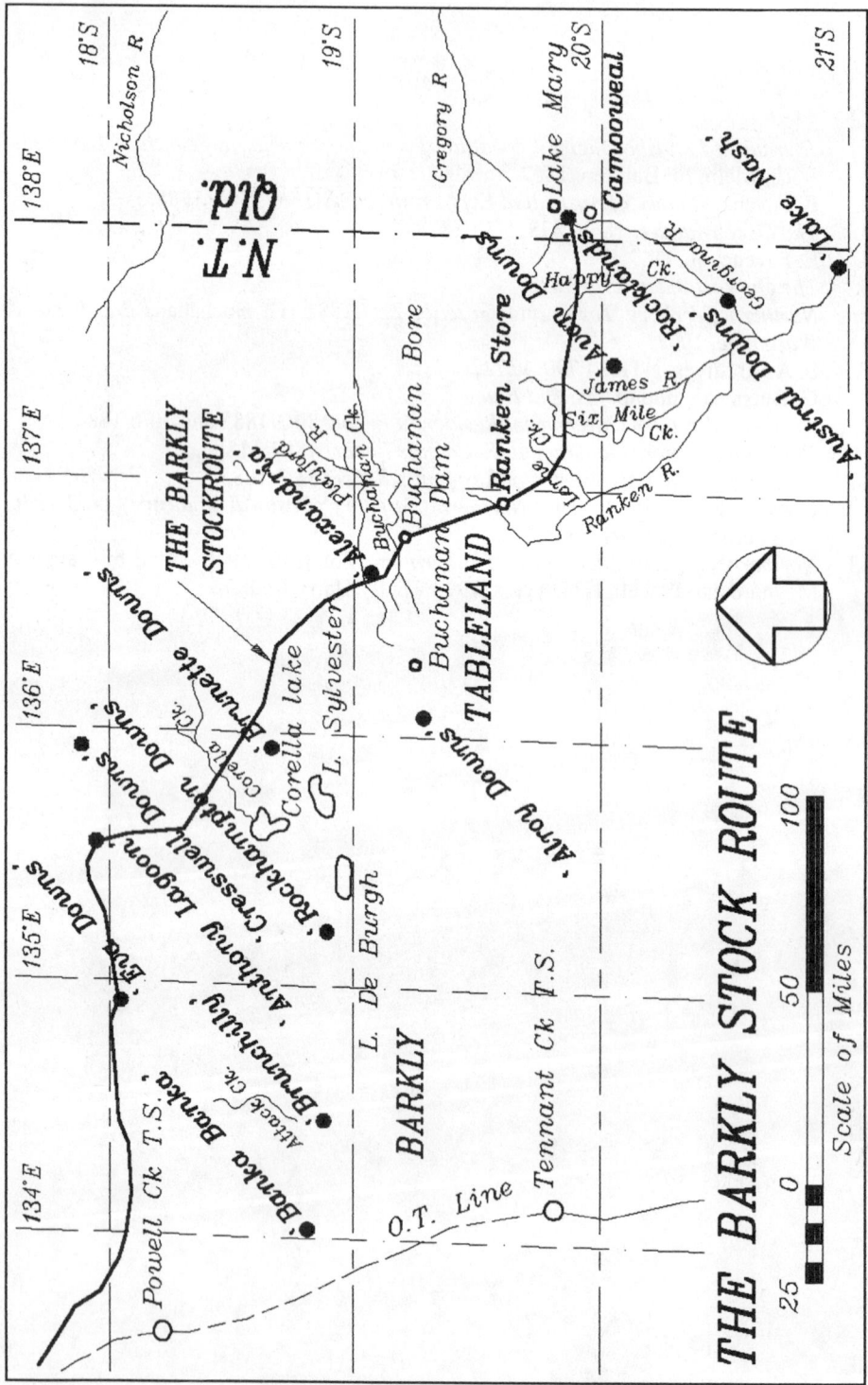

THE BARKLY STOCK ROUTE

The Barkly Tablelands

8. Aboriginal woman "Jane" in retirement at Katherine
Buchanan Collection

In 1883 Jane was a tracker for the ill-fated search party formed to investigate the alleged murder of Harry Redford and his party on the Barkly. Most of the party perished and the sole white survivor owed his life to Jane's loyalty and courage.

9. Alexandria and Avon Downs stations stock camp at Lorne Creek
John Oxley Library collection, Northern Territory Library PH 171/172

8.

The Old Coast Track

1878-1879

The Old Coast Track
Was a road of romance
Where the wild abo roamed
And the gay brolga danced.

                                    W. Linklater

The story of the Old Gulf Track, the droving route that carried the migrating herds of cattle from the eastern States of Australia to establish pastoral settlement in the Top End of the Northern Territory and the East Kimberley district of Western Australia, has all the drama and excitement of tales from the American West. Although only in regular use for a relatively short period - approximately forty years from 1879 - this track was the key to northern pastoral settlement and so is of great historical significance. Tens of thousands of head of cattle passed this way from Queensland in the 1880's to stock Territory and Kimberley leases. Then, because of lack of local markets, droughts closing alternative droving routes and the bank crashes of the early nineties, many mobs made the return journey. The track was travelled by drovers, settlers, gold-seekers, Chinese immigrants, itinerant workers and those on the run from the law. To date there appears to have been no detailed historical or archeological study of the route or documentation of significant sites. This seems incredible when the Gulf Track has been described as the Australian equivalent of America's Oregon Trail!

Some say that the Old Gulf Track started from Settlement Creek, just west of the Queensland border. Other sources nominate the jumping off point as Turn Off Lagoon on the Nicholson River, just east of the Queensland border. In any case, it wends its way along the strip of land that lies between the Barkly Plateau and the sea until the Roper River is reached. From there the passage was inland along the valley of the Roper, via Old Elsey station and Mataranka (Bitter Springs) then north up the Overland Telegraph line to Katherine. From Katherine other stock routes led north to the Darwin area, west to the Victoria River District or on to Western Australia, or sometimes south to more inland properties.

The history of the Gulf Track commenced in 1844-5 when Ludwig Leichhardt became the first European to cross this country. Leichhardt had persuaded his friends to fit out an exploring party, which he would lead, in an attempt to connect the mother colony with north Australian settlement recently formed at Port Essington on the Coburg Peninsula. He set out from the Darling Downs, then the northernmost limit of settlement, and travelled north up the eastern side of the Dividing Range, keeping about 100 miles inland and parallel to the coast where there was plenty of water. When he reached the Burdekin River Leichhardt followed it to the Lynd and then made for the shores of the Gulf. Here he came into contact with hostile Aborigines and one of his men, Gilbert, was killed. Rounding the head of the Gulf, Leichhardt encountered thick scrub, food ran short and his horses began to die, but he continued doggedly on to Port Essington, arriving there on 17th December, 1845. He named many of the Gulf rivers after members of his party.

In 1872, W. D'arcy Uhr was the the first recorded pioneer to use Leichhardt's track. An ex-policeman from Queensland, he was employed by Dillon Cox to take 400 head of cattle from near Charters Towers to the Overland Telegraph base camp at Roper Bar. Roper Bar was originally named Leichhardt's Bar because this is where Leichhardt crossed the Roper in 1845, and the landing there was called the Roper Landing. The bar is eighty miles from the mouth of the river and there was no other crossing below it.

By all accounts Uhr's trip was particularly torrid. Uhr was a good bushman but had the reputation of being a 'bit of a wild man'. It is said that he shot his way through the Aborigines and quelled unrest among his men when he stock-whipped one of their number for daring to pull a gun on him. As a policeman he was relentless in the pursuit of horse thieves and other lawless characters, but he was hot-headed and resented authority.

Records covering the very early days of the Old Gulf Track are scarce and no doubt there were many forgotten men who dared the dangers of that lonely road before it became an established route. One story which has survived is that of a man named G.D. Latour who bought 500 cattle from a property north of Charters Towers and employed William Nation to lead the drive to Pine Creek. This party ran into serious trouble with men deserting and cattle becoming lost and taken by Aborigines. The trip was plagued with tragedy from beginning to end. The men suffered from fever, were delayed by flooded rivers, dry stages and rushes. The Aborigines stalked the party and speared the cattle. It is not surprising that the men deserted in droves leaving only Nation and Elvey with the cattle. They were forced to desert the stock to try and save their own lives. They got as far as the Limmen River but couldn't cross the floodwaters so they made their way into the ranges in an attempt to find a crossing. Nation became seriously ill here and Elvey made a brave, lone journey through to Daly Waters to get help. When Elvey returned he found Nation dead, probably from starvation. He obviously anticipated death as his will was in his pocket and he had not forgotten his loyal mate. Nation Creek was subsequently named after him.

In 1877 vast tracts of land had been taken up in the Territory by applicants from other colonies. These pastoral leases had to be stocked within a certain time and the cattle to stock them came predominantly from Queensland. Some came from as far away as Victoria and New South Wales, while others came from South Australia.

The Gordons had heard that Travers and Gibson wanted to stock Glencoe Station in the Top End of the Northern Territory and were looking for an experienced and reliable drover. Gordie says that Nat had, 'long conceived an idea such as this,' so he was eager to volunteer his services. The three were not overawed by the immensity of the task they had undertaken and being seasoned bushmen, were not terribly concerned by the dangers they would have to face. To them it was just another job that had to be done. Following negotiations with Travers and Gibson, Nat was given the commission. The 1,200 cattle under his charge were to be the nucleus of a breeding herd. Glencoe, on the Margaret River south of Palmerston, had a good network of waters and well grassed river flats, with plenty of high ground for stock to escape from the wet season floods. The new station was well placed to service the Palmerston meat market and early indications regarding its potential as a grazing property were very favourable.[1] The well known Queensland graziers, Messrs Travers and Gibson leased the property in 1877. Roderick Travers and his partner Gibson owned Punjaub Station in the Gulf Country, the northern part of Aramac Station in the central west, and had pastoral interests in the Peak Downs area of Queensland.[2] It is very probable that Travers and Gibson had already made Nat's acquaintance and knew of his reputation.

The deal settled, Nat, Hugh and Wattie rode the sixty miles to Copperfield together for a family reunion. For Kate and Gordie, Nat's infrequent homecomings were a time of joy and excitement. Although Queensland boasted a reliable packhorse mail service to newly settled areas, correspondence between the couple was infrequent. Occasionally people Nat had met en route would pass word of him to Kate but mostly all the news she got was via an occasional telegram. Gordie recalled that when his mother had not heard from his father for a long time he could sense her fears despite her calm demeanour. It couldn't have been easy for Kate to bring up her son alone and manage family finances not knowing when she would receive the next news or cheque from Nat. There was no social security payment to fall back on in times of need then.

> *On occasions we were down to our last penny before the long awaited cheque arrived from some frontier outpost. One cheque received was for seven hundred pounds!*

Knowing he would be absent for a least twelve months Nat tried to spend as much time as possible with his family. He moved them from Copperfield to Ipswich which was a more central area to organise the drive. By doing this he was able to stay with Kate and Gordie and enjoy some family life while he was making preparations. Gordie fondly remembered two events from that time. His father taught him how to shoot with a pistol and also gave him the responsibility of taking delivery of horses. Demonstrations of affection between fathers and sons were not usual in those days so these two gestures would have meant a lot to the boy. Nat rarely praised his son yet neither did he blame or criticise him. He managed to impart his love in many different ways; first, by giving him responsibility and including him in his work and plans; second, by trusting him to love, respect and obey his mother during his own long absences; third, by giving him a good role model and teaching him the laws and lessons of the bush.

At fifty two years of age Nat set out on another new chapter in his eventful career. As ever, he was in the vanguard, this time leading the great exodus of cattle from Queensland to the Territory and beyond. Leichhardt's old route around the Gulf was the only option, the shorter Barkly route being too risky. Although not the first to use it, Nat was responsible for establishing the Old Coast Track and was credited with piloting in excess of twenty five thousand head of cattle over it - an all time droving record.

The droving party consisted of eight men: Nat as pilot, drovers J.H. and W.R. Gordon, W. Travers (a nephew of one of the owners of the cattle), Hume, Bright, the camp cook Charles Bridson, and an Aborigine named Harry.[3]

Nat estimated that twelve months supplies would be required to supply the party for the 1,400 mile trip. The drover's staple diet consisted of meat - fresh and then salted, supplemented by flour, rice, powdered potatoes, tea, sugar, dried apples and lime juice. The quality of the food soon deteriorated in the tropics, the sugar sweated away and the flour and rice was susceptible to water damage and the invasion of weevils. Burketown, in the Gulf Country was temporarily out of commission because of gulf fever, so supplies could not be replenished there.

Although Nat was quite fastidious in his dietary requirements, it appears that he had few qualms about using spoiled food when the need arose. Once when he was camping alone and preparing a damper for his evening meal, a visitor rode up to Nat's camp. He watched Nat with interest as he first sieved the weevils and grubs out of the flour before making a comment on its disgustingly deteriorated condition. Nat replied nonchalantly, 'It must be alright if these things can still live on it'.

Departure was from Aramac in April 1878.[4] The droving track from there and as far as Beames Brook in the Gulf Country was well known to Nat, who had pioneered it in 1864. The route lay by way of Mt. Cornish, up Landsborough Creek and on to Walkers Creek as far as Marathon Station on the Flinders River, then across to Canobie Station on the Cloncurry where a few more provisions were obtained. From here they went to Jacky's Lagoon on Talawanta and across the Leichhardt River at Floraville. After crossing Barkly Creek and Beames Brook near its junction with the Albert River, they proceeded on Leichhardt's old track across the Gregory and then up the north side of the Nicholson to Turn Off Lagoon.[5]

On the boundary of Westmoreland Station in Queensland they entered the Lesser Hell's Gate. Ahead lay new country for Nat and the success of the undertaking relied heavily on his skill as pilot, his unerring sense of direction, and his knowledge of men and cattle. From Lesser Hell's Gate the little party was in dangerous and unknown country until the cattle exited by the long rocky ravine of Hell's Gate proper and entered the smiling valley of the Roper River.

The provisions were carried in three drays which became an important factor in establishing the route for future drovers. Every dray was drawn by five draft horses with a spare horse for each team in case one went lame. Each rider had three or four horses, so with some eighteen draft and thirty stockhorses in his charge the horse hunter's job was a responsible one, demanding long hours. Up before daylight, he would search out and catch the horses belled and hobbled the night before and have them waiting in readiness for the drovers at dawn. The maintenance of the saddlery and harness in good working order was also his responsibility, and the wear and tear on these was great. All his spare time was taken up making repairs - worn girths or stirrup leathers were a recipe for disaster. The horsehunter and the cook, who had to rise early were excused from the two to three hours night watch expected of the other drovers.

The daily stages varied from five to twelve miles, depending on the terrain and the availability of suitable watering places. There was plenty of grass and topfeed but mostly poor quality. Nat was rarely out of the saddle as he daily rode ahead to find the best track, water and suitable campsites. Travelling alone, his method was to ride to the site he had marked for the next camp and then press on into new territory to mark the next night's camp, another five to twelve miles. Then he would ride back to meet the oncoming cattle. This round trip could be up to thirty six miles and he was lucky to get back before dark. Next day Harry, the Aboriginal tracker, would follow Nat's tracks and guide the cattle to the marked campsite. Once, on an overcast night when the cattle and horse bells were silent he inadvertently bypassed the camp on his return journey. After a lonely night camp he scouted around at daylight until he picked up the cattle trail, and rejoined the herd just as his concerned men were about to start a search party.

If the going was easy the drays and spare horses travelled faster than the cattle, sometimes reaching camp by midday. It was usual for each drover to carry his own lunch, but if the day's stage was short they would arrive in camp for a communal meal. Dinner camp was generally a casual affair. The men squatted on their swags under a shady tree, bolting damper and beef and drinking sweet, black tea boiled in the billycan; no plates or cutlery - except pocket knives. The cook used a big knife for the damper and a big knife and fork for the beef, and each man had his own pannikin on his saddle. When the water for the billycan was yellow and thick, epsom salts was used to clear it. Kitehawks circled lazily over the dinner camp waiting for scraps discarded by the cook and the jealous crows cawed from their vantage points in the trees.

The first encounter with Aborigines was at Redbank Creek, some thirty or forty miles east of the Calvert River. All were aware of the unseen menace because the cattle, and more especially the horses, became unusually nervous and restless. When one of the horses was speared during the night their suspicions were confirmed. Between Redbank Creek and the McArthur River, the Calvert Aborigines conducted a war of nerves and a campaign of intimidation against the drovers. On the Calvert River another horse died of spear wounds and the remaining horses were thoroughly spooked, so Nat ordered a double night watch.

Although the Aborigines did not come close during daylight, they continued their determined protest from the cliffs on the other side of the river. The new-comers were harangued by a loud barrage of shouts accompanied by threatening gestures. When camp was made some miles on, the double watch was set. During his turn on watch Nat took a shot at a shadowy figure that turned out to be a black stump. There was no moon but the starlight was brilliant, casting shadows that provoked his overactive imagination. However, tracks were found all around the camp in the morning so it wasn't all in his imagination.

The trail between the Calvert and the Robinson Rivers lay across poor country, covered in parts by thick scrub which had to be cut to make a path for the drays. These were often delayed while the track was cleared or cuttings made at difficult creek crossings. Although it was hot, exhausting work, the men did carve out a well defined route for subsequent drovers. Troubles continued. Some of the cattle were mauled by crocodiles while watering at the Robinson River and a surprise attack by Aborigines at Snake Lagoon resulted in the loss of another horse.

The Calvert Aborigines ceased their defense campaign when the McArthur River was joined fifty miles beyond Snake Lagoon. At the end of the dry season the last two stages before reaching the McArthur are usually waterless but in that year there was no problem finding ample water for the stock. The party crossed the McArthur River about forty miles inland from its mouth, just above the present township of Borroloola, and came upon the first open grasslands after 350 miles of forest and scrub. The feed was good here so Nat decided to spend a week or two spelling his men and stock. This rest allowed some newborn calves to gather strength and by the time the plant was on the move again, most of the calves were strong enough to join the herd. To avoid delay it was customary to kill the newborn calves because they couldn't keep up, but the bereaved cow then became a liability, continually trying to break away from the mob to find her dead calf. The few that were too weak to travel with the herd were carried on drays for two or three days and allowed to join their mothers at night. This created no end of trouble and plenty of fun. It was difficult to separate the calves from the protection of their mothers each morning and load them on the dray, but at the end of the stage the hungry calves and their milk-laden mothers paired up quickly.

Refreshed by the spell the party pushed on across Rosy, Batten and Nation Creeks. Good grass was found on the Limmen River, despite the fact that it was the end of October and the surrounding country had been burnt out by dry season bushfires. By this time provisions were getting dangerously low, so Nat and Wattie Gordon set out with packhorses to get supplies from Katherine, 300 miles away.

Hugh Gordon was left in charge in Nat's absence and he was familiar with Nat's rules that independent Aborigines should never be allowed into camp, and that no man should be left in camp alone when they were around. When some Aborigines who could speak Pidgin English wanted to come into camp, Hugh refused. The Aborigines disagreed and argued forcibly that they had had previous contact with white explorers and therefore eligible for entry, so Hugh reluctantly gave permission.

On 13th December 1878, Charlie Bridson rode out to look for a missing horse leaving his companion, W. Travers, alone in camp. With Nat and Wattie gone the party was shorthanded so all the remaining men were required to muster the cattle in readiness to move on to fresh feed fifteen miles west of the Limmen. When Bridson rode off, there were three Aborigines in the camp and Travers was busy mixing a small damper. He had positioned himself with a good view of the camp and had a gun in its holster on his belt so it appears that he was prepared for trouble. Travers knelt down with a bucket of water and a soda and acid bag for baking at his side, and busily kneaded the dough on a baking cloth. He did not hear the silent assassin creep up behind him and did not see the flash of the razor sharp, stolen tomahawk that killed him.[6]

When Bridson returned with the lost horse two hours later, he disturbed the Aborigines in the final stages of looting the camp. When they saw him, they took off at speed into the bush carrying the few remaining provisions and tools. Bridson took in the gruesome scene at a glance. Amid the chaos of the looted camp Traver's body lay where it had fallen, his hands were caked with dried dough and from his nearly severed neck his lifesblood spilled onto the baking cloth.

Terrified, Charlie didn't dismount to investigate the scene further, but rode at a gallop out to the cattle to alert the men. Immediately all hands set off in pursuit of the perpetrators, but although they located where the plunder had been divided among the group, their small numbers and the terrain rendered their mission impossible. The sorry band rode back to the gruesome scene of the crime. Young Travers had been a popular member of the party and a good mate and undoubtedly the men must have felt some responsibility for his death, especially poor Hugh Gordon. Isolated and vulnerable, faced with the looted camp and by the grim evidence of human mortality the future looked bleak. They were under-manned in an alien, hostile environment, their mate had been most foully murdered, all their remaining provisions had been stolen and they were even deprived of essential tools like billycans, tomahawks, shovels etc. It was fortunate that the men carried their own pannikins and pocket knives with them. In the cool of the evening Travers was reverently laid to rest and to mark his grave his name and the date engraved on a tree.

Two days later the same Aborigines were disturbed robbing Travers' grave. The drovers were so enraged that they opened fire with intent to kill, but the range was too long for the guns to be effective. The quarry was tracked into rough country but as before, because of the nature of the terrain and insufficient numbers of men, the mission became too risky and was called off.

Hugh Gordon decided to shift camp as previously intended to a creek fifteen miles to the west. At the new camp the men were miserable waiting for rations. Starvation was a very real threat since the remaining supplies had been stolen. Only tea had been overlooked by the robbers and this was a poor supplement to a beef diet. A night watch was set and for the men afflicted with hunger pangs and addicted to tobacco the long solitary nights without a smoke were unbearable. So unbearable in some cases that they resorted to smoking tea leaves.

Unaware of the drama taking place on the Limmen, Nat and Wattie were setting a cracking pace for Katherine via the Roper Bar. Here in a generous gesture, Wattie gave one of his distinctive shirts to a friendly Aborigine. After arriving in Katherine on 13th November, no time was wasted, the supplies were bought, loaded onto the packhorses and the pair immediately set out on the return journey.[7] Passing through the Elsey, Nat was fortunate to meet Sam Croker who offered to join the drive and lend a hand. Although anxious to get back to the cattle, Nat's pace was now limited by the heavily laden packhorses.

In the meantime, the Roper Aborigine wearing Wattie's shirt presented himself at the new camp. The shirt was instantly recognised as Wattie's and the men, already trigger happy because of previous events, and aware that the boss was overdue, feared the worst. Charlie Bridson was prepared to shoot first and ask questions later but was warned off by Hugh Gordon who kept a cool head and prevented further bloodshed.

The droving party's situation was very grim. The combined effects of semi-starvation, isolation, fear for their lives, the continual spearing of the cattle and concern for Nat and Wattie lowered the morale of these normally stoic frontiersmen. Hugh had to make an unenviable decision for a drover. He had to weigh up the huge investment made in the cattle by his employers and the loss of his own reputation as a drover, against the welfare of his men. His decision was to wait a couple more days for Nat before abandoning the cattle and leading his men to the Overland Telegraph line. All preparations were made to break camp and everything that was unnecessary for the journey was buried and their meagre possessions were loaded onto one dray. On the very day they were due to move out, Nat returned.

Fresh rations filled their shrunken bellys and revived them physically while their leader's return with Sam and news of the outside world lifted their spirits. Now with a full complement of men a punitive expedition was organised against the Aborigines. Near some steep hills, Wattie found a coolamon with signs of caked flour and nearby Nat discovered wings leading into a gap in the range made by the Aborigines for the purpose of spearing cattle. The quarry were eventually tracked into another gap between high cliffs and when they saw the whites they hid in a cave. Silently the men crept up on the cave and concealed themselves near its entrance and waited patiently. After many hours of vigilance, a boy about twelve years old tentatively ventured to the cave entrance to see if the coast was clear. When he found no sign of the whites he gave the 'all clear' signal to those in the cave. A very distinctive, tall warrior emerged warily and after checking around carefully, made a bolt for it. Hugh Gordon immediately recognised the big man as being the leader of the group that had frequented their camp on the Limmen and shot him dead as he ran. Well beyond the reach of law, rough justice prevailed. The ringleader was killed but the innocent parties were not punished.

The remainder of the occupants of the cave were women and youngsters and from them it was learned that many of the stolen items had been thrown into a waterhole. Two women and the boy were kept to retrieve the tools and stolen equipment, and the others were released. The women dived for the tools which were successfully retrieved, then they helped to locate some of the other gear. Nat was anxious to discover some missing papers, so he restrained the captives for the night, planning to resume the search next day but by morning the trio had escaped, freed under cover of darkness, by their friends.

Next day when the cattle were mustered and made ready to move on they were fifteen head short. The remains of four bullocks killed by Aborigines were found and a further twenty five died from spear wounds over the next few days.

After crossing the Wickham they ran into the densest scrub of the entire trip and found good use for the lately recovered axes. The scrub limited vision making it difficult for the drovers to keep the moving cattle in sight. Intersecting the Hodgson River they followed it down to the Roper and passed through Hell's

Gates, an extraordinary rocky gorge strewn with immense boulders and rocks of every possible description. The anxieties and dangers of the Gulf Road were left behind and their path now lay in a more hospitable area.

The mighty Roper River was struck near the Bar, the site of the old Overland Telegraph line depot, and Uhr's destination in 1872. The feed was good and the water plentiful as the cattle moved in a westerly direction along the south bank.

During one stage on the Roper a bad crossing in a narrow creek caused some problems for Bright who was driving the dray. Charlie Bridson, the cook and usual driver of the team, had malaria and was being carried on the dray along with several newborn calves. Bright was a good driver but there was a jib in the team that refused to shoulder its share in the stiff pulls up steep banks. Nat came across the stationary team at the creek and saw Bright inspecting the ugly crossing. The calves bleated plaintively as they did at every stop, but Bridson was too sick to notice or care. Nat, a sometimes unconventional horseman, offered to lend a hand to get the team moving.

'Here,' said the Boss to Bright, 'let me have a go at them, I'll take them across. You tie up the riding horses on the opposite bank within sight of the team.' When this was done Nat said, 'Now get a few stones ready. You take the off-side and I'll take the near, and I'll show you how to drive this team.' When Bright was ready, Nat yelled, ' Gee Boxer!' With more shouts and stones thrown from both sides, the team started down the easy slope and the stubborn old shafter was dragged along on his hind feet and buttocks to the bottom of the creek. Under the fusillade of yells and stones some momentum was gained which propelled the harried team half way up the further bank. Here the old shafter, finally getting into the collar, added the bit extra which hauled the creaking dray to safety. Bridson somehow managed to survive this mad transit but the calves were bumped off and led Nat and Bright on a merry chase before being restored to the dray. Charlie, roused so unceremoniously from his sickbed, became loudly vocal about the unorthodox style of the new teamster. 'I'll be damned if I'll ride in this dray again if the Boss is driving!'

About this time F. Gordon McKeane, Glencoe's first manager, was examining the leased country and selecting a site for the head station. The building of two thatched huts was commenced on Glencoe Creek near a semi-permanent waterhole, and a stockyard of 300 head capacity was under construction in preparation for the arrival of the cattle.

> *Mr. Travers and Gibson's mob of cattle, in charge of Mr. Buchanan, are expected at the Katherine in about five weeks. Mr. McKeane arrived on Sunday last; by 'Atjeh.' He comes to take charge, and form the station.*

> *Northern Territory Times & Gazette, 21/12/1878.*

Mount McMinn, a picturesque high tabletop mountain, lying close to the Roper about twelve miles from the landing was reached on the 4th January 1879.[8]

Here the skies opened and released a tropical downpour that signalled the opening of the wet season. The heavy rain prevented further travel with the cattle so a wet weather camp was set up on the hard slopes around the mountain. Tent flies were strung up and a few old sheets of iron from the abandoned Overland Telegraph base camp were used to rig up a galley to keep the fire dry for cooking.

The country was boggy which prevented the cattle and horses wandering too far, but they were continually tormented by mosquitoes and sandflies. To get relief from the biting insects the animals would stand belly deep in the floodwaters which resulted in some being lost in the raging torrent. Many calves were born during the long delay while the rain persisted, and they increased the size of the herd by 200 head.

Because of the delay at Mount McMinn rations again ran short, so Nat and Sam set out for Daly Waters with packhorses. They arrived on the 18th January after a difficult passage through flooded creeks and boggy country.[9] From Daly Waters Nat telegraphed a detailed report to the authorities about the circumstances of Traver's death and gave information about the progress of his party.

> It is pleasant to hear that Travers and Gibson's cattle are camped at Mount McMinn, but very sad to find that young Mr. Travers had been killed by the infernal blacks. Poor fellow! It will prove a great blow to his family and friends, and will lead them to hate the name of the Northern Territory.

Northern Territory Times & Gazette, 25/1/1879

It was two months after Nat and Sam's return to Mt. McMinn before there was sufficient break in the weather to move the cattle. During a short dry spell late in March, the mob moved west along the Roper via Strangways River and Red Lilly Lagoon. More rain at Elsey Creek caused a ten day hold up. There was no pastoral settlement at the Elsey then, it was a telegraph station only. A.C. Gregory named the Elsey after the surgeon who accompanied his expedition to the Victoria River District in 1855-56. It was not a pastoral settlement until 1881 when it was taken up by Mr. Wallace. It was well watered by the Roper River to the north and Elsey and Birdum Creeks that ran through it. Being on the stock route caused many problems for the owners because unscrupulous passing drovers picked up Elsey cattle on their way through and the great numbers of travelling stock caused a prevalence of diseases such as the dreaded redwater and pleuropneumonia.

A camp was set up twelve miles down Elsey Creek and here Hugh Gordon was laid low by a severe attack of malaria. Mr Tuckfield, the telegraph operator, treated him with the traditional Northern Territory remedies of calomel, quinine and epsom salts, and he made a good recovery. From Union Camp at Elsey, F. Lucas telegraphed a report about the conditions on the track.

> Arrived at the Union Camp on the 19th, cattle are alright - across the Elsey, about twelve miles down. Will stop there for about eight or ten days, to allow roads to dry up. ...Birdum running at Ross; but falling very fast. The Elsey has risen about one foot during the last five days.

> *Road from Ross' very good. Mr. Buchanan's cattle in splendid order;*
> *he has now over 200 calves. He leaves for Katherine tomorrow*
> *morning for rations - F Lucas. Union Camp, The Elsey, 23rd April.*
>
> *Northern Territory Times & Gazette, 24/4/1879.*

Early in May, Nat appeared again in Katherine for rations and said that the stock were travelling up the Overland Telegraph line about ten days out. The cattle continued north and at the end of May 1879, the first large mob of breeding cattle to cross from Queensland to the Top End of the Northern Territory arrived at Glencoe.

From the south, destined for Dr. Browne's Springvale station, six miles west of Katherine was a mob of sheep and cattle under the supervision of Alfred Giles. They travelled from South Australia up the Overland Telegraph line. Both Giles and Buchanan made historic journeys and arrived at their respective destinations within a few weeks of each other. It is sad that the first two pastoral enterprises in the Top End, Glencoe and Springvale, should have eventually failed.

At Glencoe the droving plant disbanded. Bridson and Bright went to the Pine Creek goldfields to try their luck while Hugh and Wattie Gordon remained on Glencoe for a year and helped to establish the new station. While there they discovered and named Ban Ban Springs, after the springs of the same name near their old home, Ban Ban Station. This indomitable pair of pioneers remained in the north of Australia until 1914.

Nat and Sam headed south east of Daly Waters to do some exploring and to inspect the Buchanan Downs country taken up in 1877.

> *This land had a healthy climate and all the stock fattening capacities*
> *of the Barkly. It was free of hostile blacks, but was not well watered*
> *and this was probably the reason that it was never stocked by him.*

Despite the absence of any record it appears likely that these two seasoned bushmen extended their explorations to the Victoria River District. It would be completely in character for them to do this especially as Nat had applied for leases, sight unseen, to the west of the Victoria River in January 1878.[10] When he returned to Port Darwin in December 1879, following his mystery tour, Nat applied for leases on four blocks south of his previous Victoria River leases.[11] This seems to indicate that he inspected the 1878 blocks with Sam and found them unsuitable and decided to relinquish them in favour of better blocks which had become available. One fact is certain, the blocks leased in 1879 were more Nat's kind of country.

Northern pastoral settlement and the period of the great cattle drives had commenced. In Nat's tracks there followed a continuous movement of men and cattle migrating from the east to stock the Territory and the East Kimberley district of Western Australia.

Once the track was established and in regular use, entrepreneurs of dubious reputation seized the opportunity to make a profit by setting up grog shops and supply depots along the route. The first was Mick Cassidy's pub at Turn Off Lagoon. The McArthur is navigable for about forty miles from its mouth giving easy access by sea, so similar establishments sprang up on the McArthur River at Borroloola. Grog merchant Billy McLeod imported his supplies by schooner and set up camp at Borroloola and soon after 'Black Jack' Reid arrived by boat with a load of Thursday Island rum. Reid's wife remained to tend the business at Borroloola while 'Black Jack' sailed on to the Roper Bar and set up a similar operation. In the absence of customs officers these dealers escaped paying duty on their wares for a considerable time. When sub-collector of customs, Alfred Searcy, visited the area in 1885, he seized Reid's vessel, 'The Good Intent' and impounded his loading.[12] He also confiscated Mrs Reid's possessions in lieu of duty owed but failed to discover her cache of money hidden in an old concertina. She appealed to his better nature for the return of the concertina which she claimed had great sentimental value and the unsuspecting gentleman obliged. Mrs. Reid went on her way very happily. It was rumoured and possibly not fact that her husband died from liberal doses of his own merchandise.

The country adjacent to the stock route was soon under lease and despite Aboriginal resistance which caused great stock losses and threatened the very existence of the white settlers, the country was settled.

Limmen River Station, adjoining Valley of Springs and owned by A.K. Holden, was stocked in 1883 by Jack 'Brumby' Clark. McArthur River Station, forty miles south of Borroloola was taken up by Amos Brothers and Broad and stocked at the end of 1884. In the same year Hodgson Downs on the Hodgson River was stocked for Mason Shepherd and Company. Calvert Downs between the sources of the Calvert and Robinson Rivers was formed by Mr. Gorman in 1885. Valley of Springs, situated above the tidal waters of the Limmen River about ninety miles north west of Boroloola was taken up by John Costello, who also held Lake Nash. He was joined by his wife and children despite the wild nature of the country and the fierce opposition of its Aboriginal population.[13]

These early settlers battled against tremendous odds and suffered from lack of markets, high establishment and running costs, and Aboriginal depredations as did most of the Territory landholders. Unlike the Barkly stations these northern properties had the benefit of a reliable rainfall and good surface water supply. The Gulf country pastoralists along with those from the Barkly encouraged the Government to set up an outlet for stock or a boiling down works at Borroloola, surveyed as a township in 1885, but their attempts also failed.

The Old Gulf Road finally became redundant when bores were put down on the Barkly Tablelands between 1918 and 1921, making the Barkly stockroute a more practicable track for cattle passing between Western Australia, the Territory and Queensland.

References

1. W. Sowden. 1882. *The Northern Territory as it is*, Adelaide.
2. Len Kingston. 1981. *Notes from a Treasured Past, of Aramac and Its People, 1867-1980*.Qld.
3. *The Queenslander*, 10/4/1879, describes W. Travers as being a nephew of Roderick Travers. Charles Bridson was the cook on this trip and not Brebner as stated in *Packhorse and Waterhole* - note reference to Bridson in Nat Buchanan's telegram of the 18/1/1879 - NTRS F790 A 3294.
4. N. MacIntyre. (nd). *Capabilities of the Gulf Country*. Unpub. ms. Mitchell Library, Sydney records the departure date from Aramac as 2nd April, 1878.
5. N. MacIntyre. (nd). *Capabilities of the Gulf Country*. Unpub. ms., Mitchell Library, Syd.
6. Telegram from N. Buchanan, dated 18/1/1879 - NTRS F 790 A3294 and N. MacIntyre op.cit.
7. *Northern Territory Times and Gazette*, 16/11/1878.
8. Cite Telegram from N. Buchanan dated 18/1/1879, NTRS F790 A 3294
9. Cite Telegram from N. Buchanan dated 18/1/1879, NTRS F790 A 3294
10. N.T. Pastoral Application, 1878.
11. N.T. Pastoral Application, 1879.
12. *Northern Territory Times and Gazette* 28/2/1885.
13. *Northern Territory Government Resident's Reports 1885 to 1890*.

10. "A Gordon for Me"
The three Gordon Brothers
From Left - W.R. (Wattie), W.G. (Willie), J.H. (Hugh)
Photograph courtesy of Margaret Wass

11. Telegraph Station - Katherine River
Northern Territory Archive Service, Foelsche Collection, F399

12. Glencoe Station (circa 1880)
Northern Territory Library, Roger Nott Collection, 217

9.

Twenty Thousand Head for the Top End

1881

Nat and his brothers-in-law were still away droving, when Kate at home in Ipswich received the sad news of her father's sudden death. John Gordon had been managing a property called Aberbuldie, near Walcha, when he was killed by a fall from his horse. Shocked and grieved by the news, Kate hurriedly packed and travelled with Gordie to Walcha to support her mother. The Gordons were a very close and loving family and at this time of mourning their burden was greater knowing that two of the boys, unaware of the situation at home, were far away facing unknown dangers.

On the 12th November 1879, Nat, accompanied by Harry, departed Darwin for Sydney on the S.S. *Ocean* and then joined Kate and Gordie in Walcha before Christmas.[1] For a few short weeks he relaxed with his family and planned his next project - buying horses to overland to South Australia.

From Jim Dwyer of Tamworth he bought over 100 saddle and heavy horses at four pounds per head. Nat decided to keep one of the mob, a dun-coloured horse called Brownie, to pull a new buggy he had ordered. Gordie with a pair of horses, one of which was Brownie, was despatched to take delivery of the brand-spanking new Abbott buggy when it arrived in Tamworth. The groom at the Bendemeer

Hotel harnessed the pair of horses to the buggy, and as he was driving them around from the stables to the front of the establishment, Brownie bolted and crashed the end of the pole into the wall of the hotel. Not satisfied with this, he then threw himself onto the pole and smashed it. Gordie and the groom strapped up the broken pole with rope and it was sufficient to hold up for the forty or so miles to Walcha. The proud owners of the buggy were on the verandah waiting to admire it when Gordie swung into the drive. Kate's sharp eyes spotted the roughly repaired pole and she exclaimed, 'Oh Gordie! What a smash! However did you do that?' Brownie instantly decided to give them a first-hand demonstration. He bolted down the small paddock, taking his equine partner, the buggy and Gordie with him. The entire outfit crashed into the stable and Brownie once again threw himself on the pole this time succeeding in smashing it beyond repair. It all happened in the space of ten seconds. 'There you are Kate,' said Nat equably, 'that's how he did it.' Brownie was never used again in harness and was given to Harry to ride as he had never bucked and was very quiet to handle. Unfortunately his next stunt proved more serious. Some weeks later at a crossing of the McDonald River he bolted with Harry for only a few yards, then threw himself down and broke the boy's leg. The unfortunate Harry spent several weeks in Tamworth Hospital and Jim Dwyer was obliged to buy the rogue horse back. This was not the end of Harry's misfortunes because a few years later he was badly mauled by a crocodile and had to spend five months in Darwin Hospital.

Some of the horses Nat had bought to overland were unbroken, so a crack roughrider named Phil Gray, from Inverey, near Manilla, was engaged to break them in and then ride with Nat to Farina, once the rail head in the mid north of South Australia. Unfortunately there are no details of this trip recorded so the route they took and the adventures along the track are left to the imagination.

Following the sale of the horses in South Australia, Nat returned by boat to Sydney where Kate and Gordie joined him. They took up temporary residence in a guest house in Macquarie Street and young Gordie was sent to Sydney Grammar School to continue his disrupted education. The owner of the guest house had very Victorian attitudes, a stickler for the proprieties, her rule was law. At mealtimes she took her place at the head of the table and seated her daughter at the foot with the fourteen guests arranged along both sides. Nat and his friend, Mr Winks, took great delight in upsetting the strict etiquette expected at mealtimes. The two stirrers sat opposite each other at the end of the table furthest from the dragon and from here they did their best to disrupt polite conversation. One bone of contention was that the guesthouse provided no mosquito nets, so Mr. Winks voiced a mild complaint about the prevalence of mosquitoes. The lady of the house curtly insisted that there were no mosquitoes to worry about, except just one or two. 'Or three' added Bluey dolefully. An ominous silence fell over the table. Mr. Winks was about to be married and on one occasion Nat was teasing him about his impending wedded state. Before long came the curt reprimand from the martinet, 'Mr. Buchanan and Mr. Winks, there is far too much levity down that end of the table!'

While staying at the Macquarie Street guest house, Nat's business adviser, William Kilgour, informed him that Maurice Lyons wanted to meet him. The fame of Nat's successful droving expedition from Aramac to Glencoe, together with his reputation as an explorer, had spread to big pastoral investors in Victoria and South Australia:

BIG SPECULATION

MacDermot Brothers and Scarr report .... that Mr. C.B. Fisher, of Melbourne, has joined Messrs. Maurice Lyons and A.W. Sergison in their extensive pastoral and agricultural properties in the Northern Territory of South Australia, and that arrangements have been made to start immediately several herds of choice cattle, numbering about 20,000 head, under the supervision of Mr. Nathaniel Buchanan.

'The Queenslander', *14/5/1881.*

C.B. Fisher and Maurice Lyons in partnership, had secured vast leases in the Victoria River district as well as Marrakai and Daly River Stations and were negotiating to take over Glencoe from Travers and Gibson. By November 1881 they had leased 34,000 square miles of pastoral country and more than 40,000 acres of freehold country in the Northern Territory, which required stocking.[2] Nat accepted the commission to overland the cattle from Southern Queensland to Glencoe - 2,000 miles.

Talking to the porter at the last minute he nearly missed the boat to Brisbane. My mother bundled him into a cab with his portmanteau from which a long white shirt sleeve dangled a flauntive farewell to a group of friends and bystanders.

In Brisbane Nat made preparations for what was to be the biggest droving trip of his long career. It may not have been the longest or most hazardous droving trip to be undertaken but the movement of 20,000 head of cattle under the supervision of one person was a record. Nat was responsible for the planning and organisation, the men, horses, equipment and provisions and for starting all the droving plants off at the right time and at appropriate intervals. He commenced this awesome task in May 1881 with his usual calm efficiency.

The cattle were gathered together in two contingents, one in the south of Queensland and the other in the north. The two contingents were to meet near the northern station of Richmond Downs. Tabletop wagons, drays, wagonettes, packsaddles, tents and flies, cooking utensils and all the other paraphernalia of a drover's camp had to be assembled. The men supplied their own saddles and swags. Although rations for some seventy or eighty men were needed for the southern contingent, the wagons were not fully loaded because stores could be bought as required as far north as Cloncurry. Some horses had to be bought but most were supplied by the same stations as the cattle - C.B. Fisher's properties.

C.H. Cheeseborough was Nat's secretary and he was of great assistance because of his understanding of stock and droving. Nat assigned William Gordon, the eldest of his wife's three brothers, to lead the southern contingent because he was an experienced cattleman and a thoroughly reliable drover. These three men collected or arranged for delivery of droving plants and men at the various stations in order to co-ordinate with the station musters. It was Willie's job to start each mob off along the road correctly spaced, and at the right time. The pace was set by the leading mob and initially this averaged eight miles per day. Sometimes the order was disrupted by rain, dry stages, vehicle breakdowns, missing horses and the occasional cattle rushes. Willie Gordon rode up and down the seventy mile long line of stock that travelled the western roads of the Warrego, Barcoo and Thomson.

Before the mobs could get started, the cattle from the various stations had to be mustered and branded, and horses broken in. The centre of this activity was St. George, in south western Queensland.

> From Mungindi to Charleville, from St. George to Thargomindah, and all along the Maranoa and Warrego rivers, the talk on the stations and at camp fires was of the impending great cattle movement right across Australia.

Gordie's description of the cutting out camps captures the essence of stock camp life in those heady days:

> Mobs from Fisher's own and neighbouring runs, Durhams, Devons, Herefords and mixed were being mobilised. GG6 and NN5 were the foremost brands. Purchased cattle were branded in a crush with the new company's brand. Some stations had bullock and heifer paddocks and these entailed short musterings. The majority however, had to be cut out by trained or special 'camp' horses from the several mustering cattle camps. Part of the mixed herds were mustered each day from a prescribed area, the draft for the road being held night and day by stockmen until the number to constitute one mob or unit had been made up. Young cattle don't travel well on long trips so all the selected cattle were over three years of age.

> And so the nuclei of these mobs was added to daily as the station musterers brought in their quotas, small or great, until the complement was reached. The musterers also did night watch if necessary and were always sent on the road for a few days in order to steady and train the mob fresh from the freedom, to an ordered grazing walk and a man encircled night camp.

> The cutting out camps were centres of furious activity and apparent confusion. Sweating horses and men were inextricably mixed with the ever circling herd, amid the dust. But it was not so. The confined body of cattle with two horsemen, cutters out riding through them, was very difficult to keep together. It required from six to ten men continually riding around the circumscribed area of the cutting out camp to prevent spreading and occasional breakaways from the bellowing, insurgent and rebellious throng.

The good camp horse knows his quarry as soon as the whip or spur is popped on him and is the embodiment of speed, dexterity, intelligence and beauty. Propping, jumping, wheeling on his hind legs, anticipating by a fraction of a second every refractory turn of a beast. All his rider has to do is sit on him - not an easy job for some, but an intensely exhilarating one. He is the complete artist and is reserved almost entirely for this work.

After selecting their beast horse and rider must be avoided by the camp-keepers as they emerge with a dash from the surging crush, dodging, fighting shouldering their way to the mob formed by those already cut out.

Sometimes a shout is heard: 'Lookout! That b....... red bull is dangerous'. A fighting bull or an old 'piker' bullock must be avoided because he will 'go right through you' if you are in his way. Seasoned stockmen need no warning. These animals are most dangerous when fighting amongst themselves, then everybody and every beast must make way for them.

In the thick clouds of dust there was a serious risk of accidents to novice or reckless stockmen. It was easy to gallop into an obscured animal. If it was a calf then horse and rider fell over it and hit the ground with terrific force. Men have been killed like that.

With cutting out complete, the mob for the drover is then watered and taken away, while the remainder are sometimes watered before being allowed to quietly disperse. These days the art of camp drafting is kept alive by the sport of the same name and provides a wonderful demonstration of the skills required of horse and rider with the added flavour of competition.

Fresh cattle had a tendency to rush at night and so extra men were pressed into service to watch them. Early in the drive the sudden rush of 1,200 bullocks caused the death of one young stockman. Paddy Fitzpatrick was trying to turn the leaders when his horse fell and he was trampled to death by the maddened throng.

Nat counted each mob as it left the stations and when the last of the first contingent was safely on its way he returned to Charleville by coach, then by train and steamer to Townsville, to marshall the northern mobs. Willie Gordon was in charge of the entire southern contingent - 16,000 head in seven mobs - as well as a mob of his own. The other boss drovers were W. 'Big Bill' Farquharson, Harry Farquharson, Jack Furnifull, Charlie Craig, Kennedy and Scott.

Leading the cavalcade was 'Big Bill' Farquharson with cattle mustered from Norley Station. The two rear mobs were all bullocks from C.B. Fisher's Wilmot and Currawilinghi Stations and they caused Willie Gordon the most worry because bullocks had a tendency to take fright and rush during the night.

The cattle would come slowly, some playfully, on to the camp, perhaps one used recently by the preceding mob. Cattle were generally quieter on a camp site that had been occupied previously: though it is argued that it makes no difference to their behaviour. Walking contentedly on to it with full bellies they showed only evidence of a satisfied home-coming whatever uneasiness might develop later.

Sudden rushes happen on the best possible camps and are mainly caused by some accidental or untoward incident, such as striking a match or the night horse shaking himself.

Paradoxically perfect stillness is not infrequently, with bullocks, a contributing cause of the most dangerous and uncontrollable kind of stampede. To minimise the disturbing effect on the mob of any sudden or unusual noise resort is often had to singing or whistling while riding round, and though often raucous and unmelodious, and cursed by light sleepers, the device is supposed to be effective - at least in the latter respect!

Willie Gordon had reason to remember one of the rushes. During first watch when the sleepy mob had settled down and everything was quiet, a hunting dingo flushed out a kangaroo rat hiding in the porcupine grass. The frightened kangaroo rat hopped right under the nose of a sleeping bullock a few yards out from the mob. With a fizzing snort the startled bullock leapt to his feet, and as if by mental telepathy the panic spread and his neighbours rose as one. The entire mob took off at top speed with the thunder of hooves, the rattle of horns and the crack of breaking saplings.

Harry Farquharson and an Aborigine named Spider were on watch and Vampire, an experienced night horse, was saddled and tied up ready at the camp in case of such an emergency. The drumming of galloping hooves woke the camp and instantly Willie Gordon was on Vampire's back, and after the rushing herd. The moon had not yet risen, in the dim starlight he could see nothing, so he left his immediate course to the sight and instinct of his well trained horse. The noise of the headlong charge and the occasional loud 'Whoa' from Harry gave him the general direction.

Bending down over Vampire's back to avoid low timber, Willie yelled to Spider to come round to him in order to force the cattle into a turning movement. He reached what he thought was the lead and joined Harry in his attempts to ring them, and then found that the inexperienced Spider had ridden up the other side of the cattle and was trying to ring them the opposite way. Not only was this extremely dangerous for the riders but it completely cancelled their efforts. In the darkness and confusion they were unable to communicate the correct method of 'ringing' to Spider and consequently he and Harry almost collided as they cut through a string of racing steers. When Harry realised the reason for their unco-ordinated action he let fly a string of oaths at the unfortunate Spider. Finally they steadied the bulk of the mob about a mile from the camp and Willie set out after the breakaways. Galloping over the open plain in the direction they had taken, Vampire signalled that he was on the right track by pricking up his ears and putting on a spurt of speed which enabled Willie to round up some of the leaders. By this time the moon had risen throwing a bit more light on the scene. When the cattle did not want to return to the herd, Willie realised that these cattle, although part of the breakaway mob, were not the leaders.

Daylight revealed a shortage of eighty head, so with fresh horses Harry and Spider set out on their tracks. It was dark before they got back to the camp with the escapees, which were greeted with welcoming bellows from their mates in the big mob.

Three riders at a time were delegated to patrol in half night shifts in an attempt to limit further rushes and to prevent a stampede over the plant and men in camp. Spider obliged by announcing to Willie, 'Me singem strong fella corroboree all right', and his raucous, full throated, staccato babble caused many complaints from the light sleepers.

Because it had been a good season Willie faced the 900 miles to Canobie on the lower Flinders, with few qualms. Dams and waterholes were full and conveniently spaced, and although there were some dry stages they were quite negotiable. The drovers in charge of the seven plants were competent and reliable, and although stockmen sometimes left the plants this created no problems as replacements were easy to find en route.

The routine of camp life was soon established. Each man spent about sixteen hours in the saddle, mainly at a slow walk behind the cattle or on the wings during the day, or circling the mob when on night watch. Boredom was relieved periodically by a rush or nervousness among the cattle which kept the night horses on their toes. Food was good and fresh vegetables were frequently obtained from Chinese or other gardens along or near the stock route. The abundant beef ration was only occasionally varied when the cook or the horsehunter took the time and trouble to shoot a bustard or other wild fowl.

The southern and northern contingents of cattle were due to join forces at Richmond in five months. When the battalions of cattle were on the blacksoil downs of the lower Thomson River, winter rain made life very uncomfortable for a few muddy stages. Men new to this country commented on the peculiarly unpleasant odour caused by the rain on the gidgea. The water slowly rose in the channel and watercourses forcing from cracks and fissures in its path all sorts of creepy crawleys, particularly snakes and centipedes. Athough only light, the rain was sufficient to stop the drays, so to avoid delay the camps were serviced by packhorses.

Eventually the leading mob approached one of the many camping grounds around the little frontier township of Arrilalah. In 1881 Arrilalah was the centre of immense cattle and sheep runs and was typical of other far western towns of the period. They all depended on a few stations and passing drovers and carriers for their existence. Situated on the banks of the Thomson River, it was a popular meeting place for stockmen, carriers, and waiting or travelling drovers. At its frequent race meetings and during the passing of stock, the town was a hive of humanity of all hues indulging in fighting, racing, gambling and drinking. The Arrilalah police had a reputation for leniency and the grog for its questionable quality.

The people of Arrilalah were accustomed to large mobs passing through heading west to stock country on the Diamantina, Georgina or Burke rivers, or north to the Gulf Country. Stock passing through destined for the Territory was new to them and so the influx of the Fisher & Lyons cattle created plenty of interest in the town.

Near Arrilahlah the broad sweep of the Thomson River narrowed and turned west by a stony ridge where it had scooped out a deep permanent waterhole which supplied the town. In the dry season when household tanks were empty, water carts drawn by horses and even by goats did a good trade. The 'green fingered' Chinese had flourishing gardens from which they supplied fresh vegetables to the townspeople and those passing through.

Below the waterhole the river course flattened out again to its half mile wide bed. The flood line was marked by lignum and bluebush and an occasional coolibah or river gum. These, and a few clumps of gidgea, bauhinia and beefwood near the northern bank were the only trees because they were being sadly depleted by drovers' and carriers' cooks and the Arrilalah inhabitants.

> *All around as far as the eye could see extended the unlimited downs. As the sun rose high this 'vision splendid' shortened its perspective by miles and a shimmering heat haze marked the rapidly closing in horizon. The magical mirage then took on the form of phantom water instead of real distant fields of grassy plain. In the early morning light, for a short time before and after sunrise, faraway stock and other objects often appeared upside down as if reflected in water.*

From Arrilalah the cattle headed north up the Darr River. It was near this river that Nat and Cornish had come across the tracks of the Burke and Wills exploring expedition of 1861. The 16,000 head southern contingent, travelling now at ten plus miles per day followed approximately Burke and Wills route to the Flinders River. They went via Kynuna and Fort Constantine Stations to join up with the newly formed second contingent that Nat had been organising in north Queensland.

References

Note: The unpublished ms. *Old Bluey* was the prime source of the information in this chapter.

1.   Darwin Shipping List of Passengers, N.T. Archives.
2.   Alfred Giles, *First Settlement in the Northern Territory.* Unpub. ms. S.A. Archives A5837.

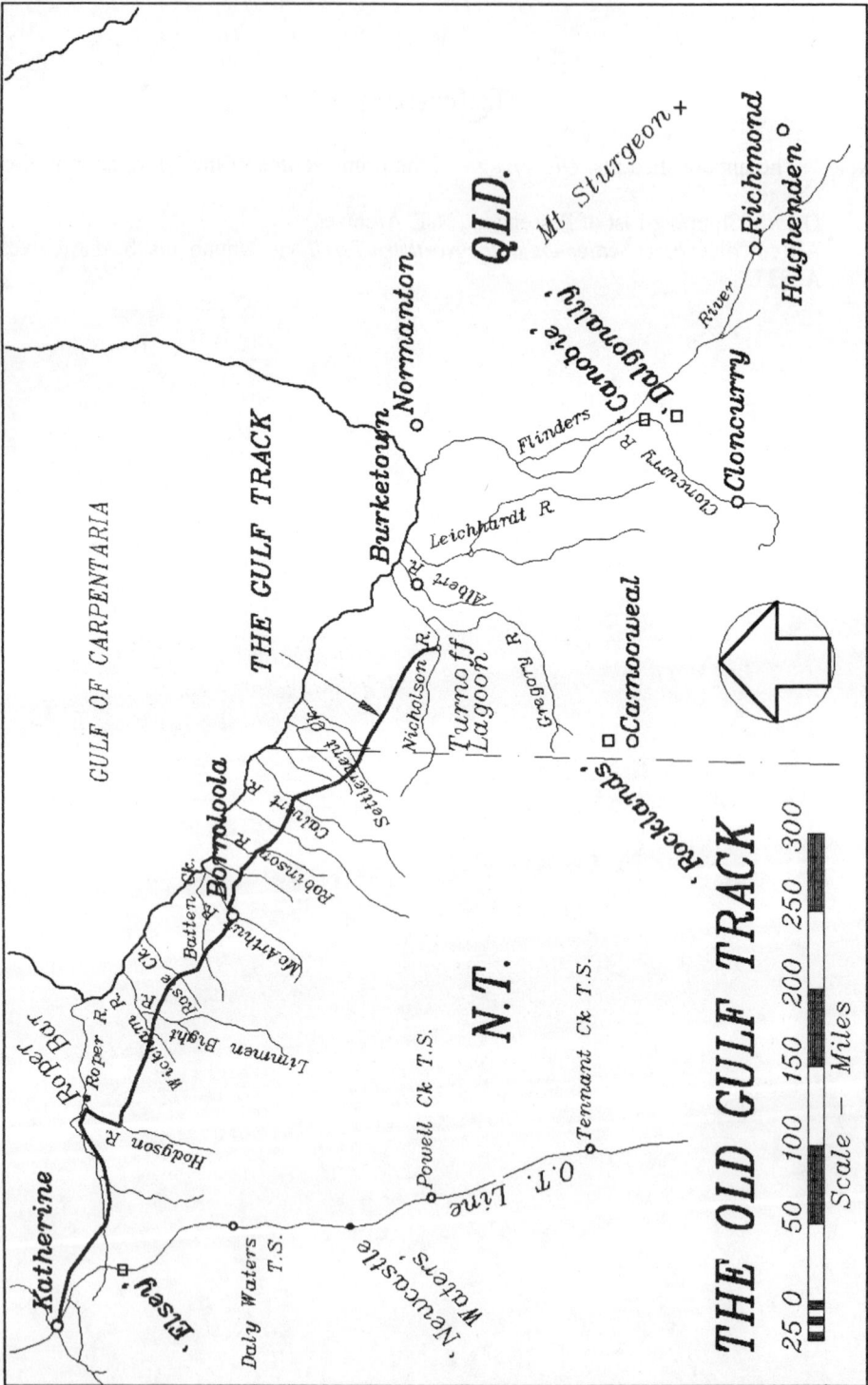

THE OLD GULF TRACK

The Old Gulf Track

13. Willie Gordon, drover in charge of 16,000 head from St. George to Richmond
Photograph courtesy of Margaret Wass

14. Richmond Downs Homestead - Drawing by Charlie Flannagan
South Australian Museum, Album titled "Drawings by an Aboriginal" No. 41 verso.

15. Daly River Cattle Station - The destination for some of Fisher & Lyons 20,000 head of cattle
Northern Territory Archives Service, Foelsche Collection, F399

10.

Richmond Downs to Glencoe

1881~82

While the southern contingent was snaking its way towards Canobie on the Flinders River, Nat was already in north Queensland. Before buying and organising the delivery of horses, droving plant and supplies he went to Richmond Downs Station to make arrangements for the second draft of 4,000 head to be mustered and collected at an appointed time.

Tom Cahill was given the job of delivering the new horses to Bundock and Hayes's Richmond Downs. Tommy was originally from Dalby in south Queensland. Nat first made his acquaintance at Fort Constantine Station, which straddled the Cloncurry River about where the town of Cloncurry now stands.[1] Impressed by Tom's methods of handling cattle Nat offered him the charge of 1,000 head. 'A blue eyed, hawk nosed, diminutive horseman full of energy, intuition and bushmanship.' was Gordie's description of Tom, who with his brother Matt, collected the 120 unbroken horses from Cunningham's Woodhouse Station on the lower Burdekin. Tom rode at the head of the exuberant mob to steady them, while Matt and an Aborigine brought up the rear with the broken horses and two or three lightly laden pack animals. At that time the road to Hughenden was unfenced and ran through heavily wooded and scrubby country.

Their first camp was a station yard thirty miles away where they could yard the horses for the night.

The men kept the horses at a brisk trot until midday when they gave them a short spell to drink and graze. Their method was to keep the horses hungry so that later on when no yards were available for the night, the horses would hang around eating rather than stray too far from camp. Unlike cattle, big mobs of horses can't be watched at night because they won't lie down, preferring to graze every hour or so. Even in open country it would have taken five or six men to hold them so wherever yards and agistment could be obtained the brothers took the opportunity to stay for a day or two to break in young horses. By doing this when they had completed their 200 mile journey, the mob was all broken in and ready to start work.

The 'Woodhouses', as the horses were called, were soon a much quieter and more manageable mob enabling them to be handled by two men. At Hughenden a wire from Nat advised Tom to meet him at Anning's Mt. Sturgeon Station, 100 miles or so to the north, so Matt and the Aborigine continued to Richmond with the 'Woodhouses.'

Meanwhile Nat took a train from Townsville to the rail terminus at Charters Towers, and then rode to meet Tom at Mt. Sturgeon. Here the two men collected a further fifty horses. They found they were of like mind about the methods of fast travelling. 'We'll give them old gooseberry,' said Nat, using one of his favourite expressions. They did, and reached Richmond before Matt with the 'Woodhouses.'

Richmond was a small village and here they bought two months provisions from Harris and Goldings, the only big store. A couple of drays were also bought locally but packsaddles, packbags and other camp gear were sent from Townsville by rail and horse team carriers. Men were signed on and were allocated to the several new droving plants.

Lindsay Crawford, who later became the first manager for Fisher and Lyons' Victoria Downs Station in the Northern Territory, was then the manager of Richmond Downs and he impressed Nat with his efficiency. There was plenty of activity on the station and Crawford had 2,000 head of cattle mustered on time and held by his own men. He had also mustered another 1,000 and put them in a yard for a night, but they rushed and broke out of the yard causing a lot of damage and extra work re-mustering them.

The boss drovers for the northern contingent were Hugh and Wattie Gordon, Tom Cahill and 'Galloping' Miller. Divided into four mobs, the cattle went stringing off down the Flinders to join the tail of the southern contingent 'forming the greatest army of men, horses and cattle under one commander that ever set out across Australia'. Nat Buchanan's skill and experience were soon to be tested again on the Gulf Track.

Nothing was too big for him. He chose the best drovers and organised
their many plants and supplies as his wide experience taught him, and
launched into the wilds with confidence and optimism. The column he
patrolled was sixty to eighty miles long.

The reinforcements from Richmond Downs now lengthened the long chain of camps and added to the difficulties of supplying rations to each unit so some reorganising had to be done. Burketown lay only twenty miles from the stock route a further 100 miles to the west, but it was a small and unreliable port and couldn't be depended upon for the immense loads of provisions required. Normanton, although 120 miles off the route, was the last town from which supplies could be obtained in sufficient quantity to feed the moving camps for the 800 miles of wilderness which lay ahead.

The drovers were getting beyond the range of droving laws which defined the length and width of a day's stage. The laws stated that twenty four hours notice had to be given to the stations along the stock routes, but the stations were fewer and the runs larger, so the drovers relaxed the rules. This occasionally led to disputes between drovers and squatters which were usually settled on the spot. The time and expense incurred by a summons to court was not worth the bother. The drovers were usually, at least technically, in the wrong and sometimes fist fights decided the issues, but mostly the stockman or manager agreed to a financial settlement.

The monsoonal rains deluged the centre mobs when they were just below the confluence of the Flinders and Cloncurry Rivers. The water spread out a mile or two wide through the channels and billabongs and delayed the heavily laden teams from Normanton which were forced to wait until the road became drier. Almost certainly the attractions of Trimbles Hotel delayed the teams further.

The country from Miltungra, via Canobie, Jacky's Lagoon, Tallawanta and Nieumeyer Valley Stations, was waterlogged for some weeks after the rivers had subsided. Of the several mobs delayed here, two or three were marooned between the channels for a few days. The more careful drovers took a path through the gidgea and ironstone ridges on the left bank early, before their drays were completely immobilised by bog.

Between Canobie and Floraville, on the west bank of the Leichhardt River, a network of swamps and scrubs was encountered which caused a lot of misery for both men and stock. This area was infested with mosquitoes, not to mention the occasional crocodile. Some relief from the biting, disease-carrying insects came at night when the men retired to sleep under the protection of box shaped mosquito nets but the wet boggy conditions made life generally uncomfortable. This sort of country was a trap for weak cattle and on the day that extra rations and two more men reached Charlie Craig's camp, several beasts needed helping out of the quagmire. Usually a 'Spanish windlass' was used for this purpose or it was done by rope. Neither of the methods was very safe for the men who were occasionally charged by an ungrateful beast.

One of the cattle was down in the mire with only its head, neck and back visible above the black mud. Futile efforts had been made to push it into deeper water to give it a chance to find a firmer landing place. Eventually a rope was passed loosely round its horns, and the obstinately resisting animal was hauled on its side up the sloping bank by two men and Spider. Once on dry ground the men quickly flung off the rope and ducked behind some thick lignum to hide from the infuriated steer. Spider ran frantically for the only available tree, a Goorlie or swamp wattle. The enraged steer saw him and charged the slender trunk a few feet below Spider's dangling legs. The sapling came down with the wide eyed Aborigine still attached and he fell in a patch of lignum with the Goorlie covering him. Wisely he lay 'doggo' until the steer trotted off to join the mob. 'Why didn't you jump on his back?' taunted Charlie Cheeseborough as he and his mate emerged laughing, from their refuge. 'What for you bin run away, you two fella,' replied Spider, enjoying the joke and reciprocating, 'You cobborn frighten longa that red bullock, him only play about!' Gordie explained that Spider was a Cunnamulla Aborigine and 'Cobborn,' meaning big one, was a New South Wales native word. Little amusing incidents like this served to lighten the spirits of the men and relieve the monotony of camp life.

At a camp at Leichhardt's Crossing the drovers went on strike demanding a pay rise.[2] After some discussion and deliberation this was agreed to, but still many men decided to leave the drive to squander their cheques in Normanton. Others too, who did not relish the dangers of the Gulf Track, took the opportunity to quit while they could. The average wage for the men at this time was from fifty shillings to three pounds per week and found for the trip from Burketown to Katherine. Most of the men who quit were from 'Big Bill' Farquharson's and Jack Furnifull's camps, but not, it seems, because they were bad bosses. Furnifull was very lenient and friendly with his men. 'Big Bill', however, was a bit eccentric. He enjoyed his own company, eating his meals alone and rarely communicating with his men other than through commands, reprimands or advice. On the positive side, 'Big Bill' was a fair man who never expected more of his men than of himself. Most of the stockmen respected him, though they often cursed his aloofness.

Now shorthanded, this was an arduous time for the remaining stockmen. They endured half night and occasional full night watches, iron rations of beef and damper, bog and general discomfort. This situation continued until the new men Nat had engaged in Normanton arrived with the horse teams loaded with rations.

Another problem the drovers had in the Gulf Country was the disappearance of horses - lost, stolen or strayed. The nature of the country made them difficult to trace and there was no time for a lengthy tracking search, because the day's stage must be completed. Being unshod meant that the tracks of the unhobbled were indistinguishable from the station horses along the way. It was a matter of urgency that the missing horses be found so Nat delegated Hugh Gordon for the job and appointed another drover to take over his mob temporarily. Hugh, a first class bushman, set out with an Aborigine and provisions for a week to scour the country for the strays. He collected a dozen or more, including a few which had obviously been stolen. He found them in a lonely spot, short hobbled and well hidden, with the drover's neck straps removed, in an attempt to disguise their origin.

The leading mobs were now approaching Burketown which is situated on an open plain on the Albert River. The town is close to the shores of the Gulf of Carpentaria and 180 miles north of Camooweal. Consisting of eight or ten iron roofed buildings it was the town and port of a sparsely populated area, a hard, hot but habitable place for the greater part of the year. This quiet little backwater became alive with activity when the infrequent Burns Philp & Company's small steamer arrived with long awaited stores for pastoralists and others in the vast hinterland. For a few hectic weeks the sailors and bushmen kept the publicans and the lone policeman busy day and night. The gaol was a log to which uncontrollables were chained or 'logged' for the night.

At Burketown the supplies were topped up and the plant was refurbished. The tucker boxes were stocked with dried fruit and vegetables, flour and rice, tea and sugar and the medicine chests with limejuice, epsom salts and quinine. Scurvy and malaria were the common illnesses on long cattle drives. Scurvy is a result of vitamin C deficiency and malaria is carried by the mosquito. The mobs moved ahead without further delay taking advantage of the lush grass and the safe watering places.

Across the Plains of Promise between Floraville and the Punjab Station, the life of the drovers and their men was easier. Normal watches and routines were resumed as new recruits replaced those who had pulled out. With replenished rations, full meals were provided and the improved diet combined with shorter hours in the saddle revived the stockmen.

Stockmen came from varied walks of life and experience and the more unusual ones were remembered in campfire yarns and over the bars of the pubs and grog shops. Tommy Bell, one of the newly signed on men was a typical 'Gulfer'. 'Gulfer' was the term used for stockmen who frequented the stockroute between Richmond and Burketown but rarely made the crossing into the Northern Territory. Tom was a character and was remembered because he had once been a sailor and punctuated his yarns with strange seafaring terminology. Soon after he signed on he told his new camp mates how he had once avoided the sack. 'I'd had enough of old Greedy, so I asked for me ticket, seized me painter (bridle) snared me crocodile (horse), put on me chair (saddle) and took water like a rat.'

Nat left the last three or four mobs to Willie Gordon to supervise and rode ahead to see the leading camp strike out from the Nicholson River, near Turn Off Lagoon - the start of the Gulf Track. Mick Cassidy's pub was the only sign of settlement at Turn Off Lagoon and was the last 'watering hole' before reaching

Katherine. A few of the newly signed on stockmen faced with the daunting prospect of several months isolation and hardship, changed their minds here and collected their pay. From that point on the grass was poorer and drier and the light crumbly soil along the river banks and flats rose in thick clouds as the mobs passed over it. Nat waited at Turn Off Lagoon until Willie Gordon came along and reported the progress of the last mobs and then again rode up the line of camps.

Once west of the Nicholson, scrub country was encountered, a miserable track of sand and yellow clay covered thickly with ti-tree, turpentine, quinine and staghorn. The clay soil supported little grass, some spinifex and was thickly studded with ant hills. Progress was slow, only about one mile an hour, and stockmen rode backwards and forwards behind the milling tailers. Visibility in the scrub was only a few yards and clouds of dust made it worse. The heat was intense and caused sweat to run in muddy rivulets down the dust powdered bodies of horses and men. With whips useless in the confines of the scrub, the men shouted and cursed at their charges as they drove them forward using stout green sticks.

It was easy for cattle to get cut off and left and the drovers had to take care that this didn't happen. They were aware that any that did get lost would probably be picked up by the following mobs or Willie Gordon's last mob. Although there was no penalty for short or late delivery the drovers' reputation was his 'beef and damper' so he guarded it jealously. It was not practical to count the mob each morning but the drovers became very skilled at remembering the distinguishing features of from twenty to 100 cattle. These cattle were differentiated from the others by unusual horn shape, scars, size, or distinctive body markings and were searched out as the drive got under way. Their presence of these 'markers' was a reliable indicator that the herd was complete.

From Settlement Creek to the Calvert the country, although thickly timbered, was not so scrubby. The sandy soil grew woollybutt, messmate, cyprus pine, ironwood, gums and pandanus. Because Farquharson and Cahill's leading camps were delayed by skirmishes with the Aborigines, much of the available feed was eaten out. This was very hard on the horses because they are more discerning feeders than cattle and don't eat top feed.

White men could rely on receiving a hostile reception from the tribes of the Calvert River region. Here the drovers were sometimes within a spear throw of the natives' mountain strongholds, hemmed in between the sea on one side and rugged ranges on the other. Harry Farquharson had three horses speared and was surprised to find that one carcass had been dismembered for meat. This told Harry that the Aborigines must have been hungry because they did not as a rule eat horse meat. From the safety of the cliffs above the river the Aborigines voiced their resentment and anger at the white intrusion. The drovers fired a few warning shots but Harry Farquharson's revolver jammed during this exchange and that night while he was cleaning it, it exploded and a bullet lodged in his knee. This was a very painful and incapacitating injury and in the absence of medical attention, the bullet had to remain where it was. For six weeks Harry was carried in the jolting dray without the benefit of pain killers or other medication. The bullet was never removed and Harry walked with a limp for the rest of his days.

Because of their hunger, a few Aborigines eventually made friendly overtures to the whites and conveyed that they wanted food. Tom Cahill was quick to cement friendly relations by giving them a load of beef. Through this positive contact the Aborigines became less timid and brought their women in too. Nat, mindful of the death of young Travers on the previous trip, strictly enforced his rule that no wild Aborigines be allowed into his stock camps.

The presence of crocodiles in the rivers caused apprehension among the men but little real trouble. Tom Cahill got the 'breeze up' when one surfaced close to him when he was riding in a couple of feet of water. In the dry season the only fresh water at most of the river crossings was just above the tidal waters. At the Robinson River, Jack Furnifull's thirsty mob struck the river where the water was salty and he lost quite a few cattle as a result of this error of judgement.

As the line of cattle wended its way inexorably through scrub and forest in the direction of the Roper Bar the chain of camps extended from the Calvert to beyond the McArthur River.

The condition of the working horses living off the coarse, sour native grasses on the track was deplorable. Because it was the dry season there was no fresh green pick to spark them up and they became skinny and dispirited. Some of them just gave up and had to be led all day. The men's rations were holding up well so some flour was spared to make johnny cakes for a few of the essential horses to give them some heart. The men fared well considering the conditions, there were the usual few cases of fever but only one fatality - a man named Sayle.

Travelling the same route but under different leadership were 5,000 cattle also for Fisher and Lyons. These mobs were level with Nat's rear camps and the drovers in charge were Jack Warby, Fred Smith and Barney Keiran.[3]

> Big bluff Jack Warby was rather contemptuous of the habit of calling all gullies and small watercourses, creeks. He rode up to Furnifull's camp early one evening to see Old Bluey, who happened to be there on his round of inspection. Warby's breezy braggadocio enlivened the evening meal.
>
> 'You're a mighty quick eater', said my father, as Warby gulped down what the cook had provided.
>
> 'You know, Buchanan, we have to be. I allow my men five minutes at breakfast. I can eat it in three minutes myself'.
>
> 'You didn't find that water you were looking for in those hills up Mountain Creek', said my father, drily.

'No, I ran up another creek - a creek, mind you, Buchanan, not a snake track! -but it was too dry, only a few small holes that wouldn't water a dingo.'

He had been boasting of improving the route by a deviation, but a masterhand had been before him. Jack Furnifull sitting on his rag of a swag - he was noted for untidiness - chuckled in his long black beard.

'Can you beat that, Boss, for hard feller talk', said Jack, as in the deepening dusk their fully accoutred and well mounted visitor disappeared tempestuously in a cloud of dust.

Furnifull hailed from Walcha and was noted as a shrewd horse dealer. He bought many horses for Nat. His untidy dress and shabby 'down at heel' appearance concealed a razor sharp mind and a ready wit.

Around the Robinson River and Snake Lagoon the country opened up and provided sweeter pastures. The better visibility made life easier for drover and horse hunter. After the brimming mirror pools of the spring fed Wearyan, where the tops of low flat banks were covered by pandanus and feather palms, the route deteriorated to the now familiar scrub and forest. At the McArthur River a detour was made to avoid the crossing, which was salty in the late dry season. Fresh water was found above the junction of Fletcher's Creek and here a couple of rest days were taken. The horses had a well earned break, a beast was killed for meat and the men were able to relax, take a 'bogie' and do their washing.

Then it was back to face more scrub, forest and sand. At the Limmen and Wickham Rivers they had another breather before pressing on through to the final obstacle - the gutta percha scrub. The narrow winding road cleared for Nat's horse drays in 1879 was used as a guide to keep the cattle on course through these dense scrubs near the Wickham River.

The drovers who escaped without scratches from the gutta percha's tough branches were lucky, for when the bark is broken a thick white glutinous sap exudes which in contact with a scratch or old sore, retards healing, sometimes for weeks or more.

Finally the jostling herds emerged from this inhospitable country. The mood of the men lightened and the cattle and horses attacked the sweeter grasses with enthusiasm.

Between Settlement Creek near the Queensland border and The Elsey - 400 miles -there were no stocked stations. Droving conditions over the 100 miles from the Roper Bar to the Elsey were ideal. The country was open, well watered and supported abundant dry feed. The Roper River was very picturesque, along its margins grew a variety of water-loving trees including Leichhardt Pine, fig trees and pandanus. However, it was avoided by the droving camps because of its dangerous, deceptively low banks above very deep water. McMinn's Bluff and Crescent Lagoon were passed in normal stages, but the 'Red Lily' was given a wide berth at night because it was the headquarters of all the mosquitoes in the Roper valley. Even during the day clouds of the exceptionally large and ferocious blood suckers would rise when a horseman rode to its boggy edge. Cornelius, Warby's cook, was fascinated by them as scores settled on his bare arms. 'What kind of bird is this? Must be a new kind. This country is full of bloody surprises!'

Red Lily Lagoon got its name from the flaming fringe of deep scarlet blooms, their bases touched with gold, which grew there in abundance. Unlike the common water lily, which occurs on nearly all lagoons and still water, the red lily was very rare. It grew on the edges of calm waters to a height of two to three feet and resembled a canna. Gordie only recalled seeing them in two other places, Wave Hill and Hodgson Downs.

The mobs of cattle moved inexorably closer to their destination.

> *Warby passed Elsey yesterday with 1616 head of cattle. Also a mob of Buchanan's cattle. Both mobs for Glencoe.*
>
> *Northern Territory Times & Gazette, 16/12/1882.*

The extraordinary group of deep hot springs near what is now the popular tourist town of Mataranka was passed. Day after day around noon, as Katherine was approached, big cumulus rain filled clouds - 'woolpacks' - gathered in the north west, increasing in size and maturity as the afternoon wore on.

Nat arrived at Katherine early in October 1882 with the first mob, which was W.H. Farquharson's. He waited there to see the last mobs cross the river and onto the final leg of their journey.[4] Over the final 120 miles from Abraham's Billabong to Glencoce, Nat's supervisory job was lighter because the mobs could follow a well defined road along the telegraph line. By this stage too, rations were easily obtained and some of the drays and teams were no longer required. Supplies, pack saddles and stock saddles were ferried over the Katherine River in the Overland Telegraph line boat and Bob Murray, the Postmaster at Katherine, kept Nat informed of events likely to need his attention. Storms delayed the passage of the remaining drays across the river, but only for a day or two. Once all the mobs were safely across, Nat rode through to Glencoe to oversee delivery of the mobs.

By the second week in December thunder showers had fallen up the Katherine River and it was a half banker when Tom Cahill and Jack Warby arrived simultaneously on its steep bank. Rivalry was fierce between the two men and each was determined to beat the other to Glencoe. Warby was first to attempt to cross at the only ford on the river. Hoping to gain time, he omitted the time honoured precaution of first sending some coachers to the other bank and forced the whole mob into the fast flowing water. Though flanked on both sides by horsemen, upon reaching midstream the leaders turned back upon the following herd causing absolute chaos. The resultant crush of desperate, circling, swimming cattle was swept downstream by the current. Some were drowned, a few reached the far bank, but the majority were carried as far as a half mile down stream before emerging on the same side of the river from which they entered. Warby was fortunate there was no loss of human life. One stockman caught in the swimming melee of cattle, had a close call when his horse drowned, leaving him to swim for his life. It was after sundown before Warby and his men managed to muster the widely dispersed cattle into a mob.

Tom Cahill was more careful, and first drove a few bullocks across and held them on the far bank in view of the main mob, to encourage them across. He made a good safe crossing and was now in front of Warby.

First to arrive at Glencoe was 'Big Bill' Farquharson and another drover and their cattle were sent out to leases on the Daly River. The incident at the Katherine River made Warby even more determined to beat Cahill. Over the remaining seventy miles he pushed his men and cattle relentlessly, the outcome being that the feuding pair arrived at Glencoe together. The stockmen were eagerly anticipating the final conflict. This was the scene at the Glencoe yards:

> *The park-like landscape surrounding the big lagoon on which the head station and yards were situated was partly occupied by the rival droves waiting to be counted through the yards on delivery, an unnecessary proceeding, but insisted on by the manager, Tom Nelson.*
>
> *Cahill's cattle were the nearer, but Warby met him between the wings as he (Cahill) was opening the big double gates, and with blustering challenge claimed priority.*
>
> *'My lot is going through first, Tom. Get out of the way'.*
>
> *But Tommy, five feet two in his socks, stood his ground. His burly rival threatened to knock him down.*
>
> *Tommy's blue eyes flashed as he picked up a stout but supple yarding stick.*
>
> *'Come another step closer, Warby, you b... big bully and I'll lay you out. If this won't stop you, I have something here that will!' putting his hand on his revolver. Warby's was on his saddle some hundred yards away. Bluff or not, that staggered Warby.*
>
> *'Go on then, yard 'em up you cantankerous little fat b... I'll summons you for this!'*
>
> *Cahill beckoned to his men to come on. They had watched the dispute with lively interest.*
>
> *'Well, I'll be damned if the little feller hasn't got him bluffed,' said Billy Haynes.*

As the mobs came stringing into Glencoe they were yarded and counted, and then dispatched to the other Fisher & Lyons leases in the Top End. The cattle fared badly on their new pastures because of the poor nutrient value of the grasses and the rapid decrease in herd numbers prompted the owners to move most of the cattle from Glencoe two years later. Drover Willy Gordon moved 6,000 head to Fisher & Lyons' unstocked leases on the Victoria River. H.W.H. Stevens, general manager for Fisher & Lyons, reported that the first cattle he knew of to show tick or redwater fever were in the mobs that came along the Roper in the dry of 1882. Out of the first 1,700 delivered to Glencoe, 400 died on delivery.[5] This was the beginning of the scourge of redwater fever that was to plague travelling stock in the Territory and retard the growth of the pastoral industry.

Once the cattle were delivered and the droving plants and men were paid off and disbanded, Nat returned from Darwin to Sydney by sea. It had been eighteen months since he had parted from his wife and son.

References

1. *Hoofs and Horns*, June 1949, re Tom Cahill
2. Gordon Buchanan says in *Packhorse and Waterhole* p.62 and in his article in the *Sydney Stock and Station Journal* dated 3/2/22, that the drovers strike took place at Leichhardt's Crossing and in ms. of *Old Bluey* he says it took place earlier between Canobie and Floraville. I have selected the former for this history as the leading camps could well have been at Leichhardt's Crossing whilst the last camps could have been anywhere between Canobie and Floraville.
3. *The Queenslander* 17/6/1882.
4. Register of Census returns, Cattle Movements, CRS F108 Item (1), Alfred Giles, unpublished ms., *The First Pastoral Settlement in the N.T.*, S.A. Archives A 5837
5. APP, *Government Residents Report on the Northern Territory for the year 1888.* NARU DU 392. N76 1888.

11.

First Cattle to the East Kimberley

1883-1884

Nat Buchanan again made droving history on 22nd June 1884, when he delivered cattle to stock the first station in the East Kimberley region of Western Australia.[1] The mob was made up of 4,000 head of shorthorns from Avington Station near Blackall and 1,000 head of Herefords from Beaufort Station on the Belyando River.[2] Nat took over responsibility for the drive at Richmond in May 1883, after the two previous incumbents had been sacked. From Richmond he followed his earlier routes as far as Katherine then he headed south west across 450 miles of virgin country, to establish a new stockroute. Nat's son Gordie, now eighteen years of age and having just completed his schooling, realised his boyhood dream of riding with his father.

W.H. Osmand and J.A. Panton took up leases in the Ord River country in response to enthusiastic reports in Eastern States newspapers of Alexander Forrest's explorations in the area in 1879. Forrest had led an expedition from the De Grey River in Western Australia to the Overland Telegraph Line, and discovered valuable pastoral land in the Kimberley district. The area leased by Osmond and Panton was approximately 12,000 square miles, stretching from the Negri River to the lower Sturt Creek. W.H. Osmand, a millionaire nicknamed 'Smelter' Bill or 'Billy the Smelter' by the people of Stawell in Victoria, was the senior partner in the enterprise. His partner, Joseph Panton, had risen through the ranks of the Public Service to become Chief Magistrate of Melbourne. Panton was a member of the Royal Geographical Society of Australia and a man of diverse interests which included art, music, sailing, grape growing and wine making.[3]

'Smelter Bill' had trouble finding a suitably experienced drover to supervise the cattle drive. The first leader was sacked at Bogantungan when his extravagances offended Osmand's miserly nature. He was replaced by Arthur Longdon, a twenty two year old friend of Panton. Longdon arrived in Bogantungan from Rockhampton complete with men and 100 head of horses, just in time for the local race meeting. At the races he was so impressed by the riding ability of a young man called Donald Swan that he invited him to join the drive as horsehunter. Swan was one of the few men to complete the drive to Western Australia and was to become a Kimberley identity.[4]

Longdon's style of leadership was unique.

> Longdon was a fine big boyish fellow ready for any adventure, whether offering on the road, in the infrequent towns of central and western central Queensland en route, or later with blacks and crocodiles. He usually made one of these small towns his headquarters until all the mobs had got well past. Then in his fast tandem dogcart he would overtake and visit the camps, drop a few extras by way of supplies and his flying visit completed, dash on to the next town, where he could be sure of a welcome.

Travelling as they were through settled areas, this unusual method of leadership did not adversely affect the stock. Jack (J.R.) Skuthorpe was pilot and there were experienced and reliable drovers to set the pace and direction of the drive. Jack Skuthorpe was a member of the famous buckjumping family and was nicknamed 'The Relics'. On a trip to the Northern Territory he once claimed to have found Leichhardt's journal and some human bones belonging to the lost exploring expedition. Despite stirring up much interest he never produced the famous relics and they were never heard of again.[5]

Skuthorpe was a bit sour when he got the news that Nat was to take over from Longdon. Up until now he had had an easy ride, receiving seven pounds a week to pilot a mob over well used stockroutes. Gordon Buchanan recalled:

> At Richmond, anticipating the sack, his first words on meeting my father were, 'Well Buchanan, I suppose you won't want me as pilot now.'
>
> 'No, I reckon I can find my way across Australia.'
>
> 'You'll need a block and tackle to cross the Leopold Ranges', said 'The Relics'.
>
> 'No, you're wrong Skuthorpe, the Leopolds are nearly two hundred miles beyond the Ord where I plan to strike it'.

When Nat took over command he inspected all the plant and equipment and found a stock of rum hidden among the supplies on the drays. Much to the chagrin of the stockmen he tipped the contents of the demi-johns onto the thirsty black soil plains.[6] Nat knew that grog caused trouble in cattle camps so he never tolerated it.

Gordie was assigned to Walter King's camp to serve his apprenticeship as a drover and he vividly recollects his first night in stock camp.

It was calm and cold and the night was bright with stars as Gordie absorbed the night mood of the camp. From the timber down by the river he could hear the monotonous call of a mopoke and the distant sounds of horse and bullock bells. One hundred yards to the south of the camp, the recumbent cattle rested in a dark mass in the starlight as the first watch was whistling round them. After the evening meal, the talk around the fire drifted into a discussion of bush exploits. The knowing audience of old hands started to drift towards their swags when Jack McConnell or 'Happy Jack', the horsehunter, started to tell yarns about his horsemanship. Jack continued with his stories. 'Look here, young Bluey, I stuck to that bit of forked lightning like treacle to a blanket.'

Eventually his endless flow of anecdotes was interrupted when Bill Weldon, one of the men on watch, rode quietly into camp. 'Better turn in, young Bluey. Jack will keep you up all night with his skiting.' Bill, unperturbed by Jack's unparliamentary response, helped himself to a drink of black tea from the communal pint pot kept on the lug of the bucket which stood by the fire all night. When he had finished he lit his pipe with a live coal and mounted his horse and rode off to resume his watch. While they were having breakfast in the first dim light of dawn, Jack was down on the floodbanks of the Flinders rounding up the horses. They could hear him cursing the drover's choice of camp every time his horse stumbled into the holes hidden by the long grass.

At Richmond, 'Bullocky Dick' the bullock team driver, went on a bender and half of his team went missing. Nat eventually found him prostrate from the combined effects of bad grog and dysentery and completely incapable of finding the lost bullocks, so he put Gordie on their trail. 'Don't come back without those bullocks. You can stay the night with Crawford on Richmond Downs if you are late.' These were Nat's parting words to his son.

Gordie set out optimistically, thinking the bullocks would be easy to find. He searched the north side of the river through scrub and timber and kept his ears open for the sound of 'condamine' bells, which would give some direction to his search. Richmond Downs was a cattle run so it was no use looking for tracks. He inspected several mobs of big bullocks but decided they were too wild to be workers and they didn't have the OP brand on them. Lunchtime found him at Charcoal Creek, so he rested and ate his beef and damper. By late afternoon, after an unsuccessful search up the creek he returned to the river, arriving after dark. It was a chilly night and Gordie was relieved to find refuge in a carrier's camp. The night got colder and colder and the carriers didn't have any spare blankets so he wrapped himself in a tarpaulin, and spent a most miserable night. In the morning after a strong cup of tea, he continued searching the bush all the way back to Richmond, without success.

The lost bullocks were never recovered and neither were the two stallions and some missing horses. The changeover of such a big plant to a new command was an opportunity for horse and cattle thieves. Many thought the practice of 'lifting' a necessary evil and often bushmen shut their eyes to the practice. They rationalised that it would teach the squatters to put on more men and brand up their stock. It was not uncommon for itinerant thieves to attach themselves to a station or droving camp and when the mood took them, leave with whatever stock they could conveniently lay their hands on.

Gordie's next assignment was to take over the accounts from Longdon's book-keeper at the bank in Hughenden. No doubt his father thought all that schooling shouldn't go to waste. The round trip from Richmond to Hughenden took about eight days riding, so by the time he returned to meet Nat, the cattle and droving plants had moved on. After attending to some final business matters father and son rode to overtake the cattle, which were now passing Manfred Downs sheep station.

At Miller's Lagoon, beyond Floraville, John Brodie rode into camp done up handsomely in a bright red shirt and white moles. Brodie had an interest in Cresswell Downs on the Barkly and he had been doing some private exploring up the Nicholson River. His entourage consisted of a white man and a blackboy, plus a fine plant of saddle and packhorses. The Brodie family and the Buchanan family were well known to each other.

'Well, Buchanan,' said John, as they were shaking hands, 'I passed two of your mobs today, the leading one near Harris' Lake. There'll soon be no cattle left in Queensland if you keep on taking them away like this!'

'All the better,' replied Nat. 'More room for sheep.' After his experiences on Bowen Downs and Craven, Nat didn't have much time for sheep.

The mobs were spaced over forty to fifty miles of road and because of the good season the 325 miles from Richmond to Burketown was without a hitch. Initially the drovers were Johnny Fitzroy, who was typical of the ideal drover, tall and straight and an excellent horseman. Joe McMaugh, heavy set, but a fine horseman, W.H. Young and Walter King. Changes had to be made at Burketown following a drover's strike for higher wages. As the wages were considered to be above the average usually paid, no increase was approved. This resulted in King and Fitzroy's resignation. Tragically, Young was killed by a bolting horse, leaving McMaugh as the only boss drover for the four mobs. Some cattle had been lost in the bad lands between the Nicholson River and Floraville, so the four mobs were reduced to three larger ones, enabling Ford and Weldon to take over King's and Fitzroy's camps. Bill Weldon had joined the expedition at Bogantungan and was second in charge to Fitzroy. He came originally from the Dandenongs in Victoria and was rumoured to have been a 'telegraph boy' for the Kelly gang.[7] Ford, had been second in charge to King and he was very pleased with his promotion because he was an ambitious man who enjoyed the mantle of authority.

G.R. Hedley, who had been with Fisher & Lyons cattle on the previous trip, was employed by Nat to take over the fourth plant. He was required to remain and muster the stragglers before continuing with them to the Ord. Hedley had served with Favenc and Briggs' Queenslander Exploring Expedition in 1878. On his previous trip along the Old Gulf Track he shot an Aborigine. One night when he was lying awake in his swag under a mosquito net an assassin crept up. The stealthy Aborigine lifted the corner of the net, took aim with his spear and then dropped down dead on receipt of a well directed shot at close range from Hedley's revolver. Years later this competent bushman became the manager of Ord River Station.

The parties camped on the Gregory River and reprovisioned at Burketown. The Old Gulf Road was busy that year. Immediately ahead of Buchanan's three camps were Tom Cahill and Hugh Gordon with a mob of 1,200 heifers bought from Donald McIntyre. They were en route to stock Nat's Victoria River leases. Ahead of Cahill and Gordon was 'Brumby' Clarke with a mob of cattle owned by Holden of Limmen River Station and both mobs were preceded by D'arcy Uhr with 2,000 heifers from Gooyea Station. These were destined to be the first stock on Newcastle Waters. Uhr pioneered a new route to Newcastle Waters. He followed the coast track as far as Strangways River, then went inland until he eventually reached the Overland Telegraph line at Daly Waters.[8]

On the McArthur River in early October, misfortune struck. Pleuro pneumonia broke out in the Wave Hill mob causing the death of many cattle and a lengthy three month hold up. To contain the outbreak the cattle had to be inoculated, so yards and a crush were built for this purpose. With only axes and tomahawks as tools building the yards was a big job. Pleuro pneumonia had appeared in all parts of the colony of Queensland and was characterised by acute inflammation of the chest, discharge from the eyes and nostrils, and loss of strength. The cattle were infectious for four to five days and usually died within twenty hours of showing symptoms. The disease was known to devastate teams which frequented the same resting places.[9]

Rowley Edkins was one of the first in Queensland to inoculate cattle against pleuro and it proved to be the only protection against the disease.[10] A diseased beast was killed and the blood collected, then strips of wool were soaked in the infected blood and introduced into the fleshy part of the tail of a healthy beast. This produced the desired immunity and was the method used to inoculate the Ord River cattle.

In Ford's camp one evening the cook and the horsehunter had a 'blue' which escalated to a fight. The pair were unevenly matched. Jack, the horsehunter was a big muscular chap and his high pitched voice and English accent belied his physical strength. The cook was only slightly built which gave the big man an unfair advantage. Jack knocked his opponent down and after delivering a couple more punches he proceeded to sink in his teeth. Nat couldn't believe his eyes when he rode into camp just as Jack was biting off chunks of the cook. He broke up the scuffle and disgusted by Jack's unconventional tactics he remarked, 'Why Jack, you're more like an alligator than a man.' The horsehunter became 'Alligator' Jack from that time on, but he did not bear his new name proudly.

On Fletcher's Creek, between the track and the sea there was an area of burnt out country that had flourished in response to a late October storm. This provided good green feed which was just what Nat was looking for to revive a few hundred of the weakest cattle and horses before continuing the long drive. Although it meant backtracking with some stock to Fletcher's Creek the master drover didn't hesitate because he wanted the cattle in good condition for the long road that still lay ahead. During this quiet period while the cattle were recuperating, Nat made scouting trips to the surrounding country, relocating some flammable shale which he had come across in 1879. When it was tested in the fire, it produced a jet of flame which lasted several seconds. Gordie went along on some of these scouting trips to look for camps and suitable watering places and he found his father as diligent as ever about taking the precaution of moving on after the evening meal and camping without a fire. Later they heard from a friendly Aborigine from McArthur River station that a party of Aborigines had been stalking them and this precaution had saved their lives.

Around this time a man named Fraser, and one of his two Aborigines were speared to death, on the Gulf Track. Fraser, whose destination was the Pine Creek goldfields, was working his passage with Brumby Clark. Tom Cahill, who was ahead of his mob looking for a night camp saw Fraser and told him to shift camp because it was dangerously close to the high banks of a creek. The man must have been a fool because he ignored Tom's advice. Fraser, with Clark's packs and two Aborigines, was resting unarmed, his revolver on his saddle when five or six armed Aborigines rushed yelling from the nearby creek launching a rain of spears. Fraser made a run for it but was speared first. One of his Aborigines, although wounded, survived to tell the tale. The bodies were found and buried by Nat's party.[11]

Up ahead, Aborigines had speared and scattered D'arcy Uhr's horses. Intent on retribution he formed a search party and arrived at Nat's camp just as the men returned from an unsuccessful hunt for Fraser's murderers. The suspected miscreants had been tracked into a rugged area but were alerted by the accidental discharge of a bullet from Ford's revolver, which neatly parted his beard! The stockmen were pleased to hand over their mission to Uhr and his two trackers who were, as usual, successful.

Rations began to run low because of the delay caused by the disease and the impassable dry stretch ahead, so Nat rationalised his camps from three to two. Ford and Weldon and many of their men were paid off and they made their way back to Burketown. The stocking of the Northern Territory leases was proceeding at fever pitch, so the discharged men would have soon found work with other drovers. It was decided that the mob of Wave Hill heifers should travel in convoy with the Ord River mob as far as the Victoria River District. At this time there were two camps, one on the Fletcher and another on the McArthur. The cattle had limited freedom during this extended period of rest. Apart from being boundary ridden each day to stop them straying too far, they were unattended, hence the men's workload was minimal and they had time for other diversions.

After Ford left, Gordie was reassigned to Joe McMaugh's camp. Scouting along the upper Fletcher one day George Maunders and Gordie struck some strange fresh cattle tracks. They followed them until they caught sight of a mob of about thirty, to the windward, in a thick ti-tree forest. One big micky sniffed the air, saw the men and the mob took off. The riders were after them in a flash and although their horses were in poor condition, with a bit of scrub-bashing they caught up and then let them run for a bit before nursing them to a standstill. Gordie was soon able to explain their success:

> Nearly half of them were branded cattle lost from previous mobs - mostly Fisher and Lyons' - and to them we owed our easy capture for they had not forgotten their training.

Once the mob was safely back at camp, Joe McMaugh shot a big fat cleanskin heifer and the camp feasted on fresh tender beef for a day or two. Salt was collected from a natural salt pan near the mouth of the Fletcher, and the remainder of the beast was dry salted. The stacks of freshly salted beef were then turned night and morning until all the brine had run off, then it was put into bags. The salt beef was appetising for the first week or so before it got too dry.

There were a lot of wild horses in the area, sired by near-thoroughbreds lost by Cox back in the mid-seventies. Joe McMaugh, George Maunders and Gordie decided to do some brumby-running. After a death-defying but exhilarating ride they yarded a mob of about twenty with the help of coachers. When Gordie was approaching the trap yard he saw a handsome bay colt about to beat him and pass the hidden wing of the yard. A revolver shot aimed in front of the colt turned it onto the right line, and he was caught. In the same mob they discovered a quiet old thoroughbred stallion bearing Cox's brand and Crackbell which became the best camp horse on Ord River station. These animals were a cut above the usual weedy brumbies and would bring a good price once McMaugh, who was made responsible for them, had broken them in.

By mid-December the rations were almost gone and there had only been a few isolated storms. Nat located water twenty miles ahead in Batten Creek and decided to move the cattle on. They were mustered into two camps on Christmas Eve, the lead camp under Cahill and McMaugh and the rear camp under Nat. Gordie, along with Donald Swan, Ted Lenehan, Jack McConnell and George Maunders were reassigned to Nat's camp.

Bob Button, the first manager of Ord River station, joined the drive at Christmas. A noted roughrider, brilliant horseman and a great natural bushman, Button first came to Queensland as an employee of Tozers, graziers of Warrnambool, Victoria. He had a weakness for drink and one of his eccentricities was to never leave his nine inch colt off his belt, but he made a great contribution to opening up the Kimberley. He explored and prospected extensively in the area and became one of Osmand and Panton's longest serving managers.

To celebrate Christmas Day the cook used up the last of the luxury items from the meagre supplies, and produced plum duffs and brownies. The Christmas pudding mysteriously disappeared on Christmas Eve, taken by those on night watch! Next day Bob Button and Joe McMaugh, with a string of packhorses, a number of spare horses and ninety newly broken brumbies, were despatched to Katherine. They arrived at the Elsey on 21st January 1884, but it was six weeks before they returned with the rations. In the meantime the men in camp existed mainly on a diet of tea and beef. There was precious little flour and even less sugar, so rations were supplemented with pigweed and scurvy grass, a soft leafed plant with a blue flower. Sufficient rice and preserved potatoes for one meal was rationed out on a weekly basis. In addition there was still some ship's limejuice left to help avoid scurvy. One stockman got so skinny that he had to use a small pillow to cushion his discomfort in the saddle.

Nat volunteered his services as camp cook. This was probably because with so few rations left the menu was simple. With the assistance of Don Swan, he boiled the beef with an occasional bush vegetable at night and cooked six tiny johnny cakes each morning. Johnny cakes were a luxury. Only about the size of a Sao biscuit, they could be consumed in a few mouthfuls. Flour to make the cakes was rationed to half a pint per man and all the flour was combined in one dish. Inevitably when cooked, these bush delicacies varied in size, so a fair method of distribution was found. One of the men would stand with his back to the cook and announce the name of each stockman and the cook would then pass him the first cake that came to hand. Some of the men made theirs last by only taking a few bites at breakfast and keeping the rest until later, while others couldn't resist eating the lot at once.

Travelling as two mobs during the day and combining into one big mob at night, they moved on quickly to take advantage of a few scattered storms. There was green feed at frequent intervals and the cattle, now fresher for their spell, chased each other for the sweet new shoots and tips of top feed. For five weeks they averaged sixteen miles a day, and although there was good stock feed the drovers existed on iron rations. Now and then the drays were delayed by bog, which meant the stockmen had to survive on'staggering bob' until the cook and the drays caught up.

Despite poor rations and long days, morale remained high in the two camps:

Jack McConnell was as talkative as ever while absorbing quarts of tea at the evening meal. 'Where do you put it, Jack?' enquired Ted Lenehan. 'Well Ted, you know it just seems to dry up inside of me.' drawled Jack. 'It's not like you then?' said Ted. We had all become thin and dried up, though it was Jack's normal physique.

When not on first watch, George Maunders would on odd occasions give us a recitation, which he loved to deliver posing on a log or the tailboard of the dray. Neither George or Ted Lenehan were great talkers but Ted had the longest and weirdest vocabulary of swear words that I ever heard. We were all surprised to hear that when he returned to Rockhampton he joined the Salvation Army! Jack, George and Ted only went as far as Katherine. I found the trio cheerful and helpful comrades on that, my first and hardest droving trip.

Bob Button met the drive with fresh provisions at McMinn's Bluff on the Roper, and everyone overindulged which resulted in upset stomachs all round. Almost as welcome as the food were the out-of-date newspapers Button delivered. The men, hungry for news of the outside world, read the papers from cover to cover several times over, including news about themselves:

Mr. Buchanan arrived at the Elsey on the 11th inst., in charge of 2,500 head of cattle, 90 bulls, and 160 horses, en route for the Western Australian border.

Northern Territory Times & Gazette, 16/2/1884.

By this time the Wave Hill mob was travelling independently, three days behind the Ord River cattle. Nat's friend, the officer in charge of the telegraph station, put on a grand welcome when they arrived in Katherine. He organised an impromptu race meeting and there was stiff competition between the stock horses and the telegraph station horses. The men had a chance to relax and have some fun before the next stage of the drive. There was the usual change of hands at this stopover. Joe McMaugh left the drive to join Osmand on his vessel, the 'Cushie Doo' at Port Darwin. Osmand was planning to investigate the Cambridge Gulf to see if it would make a better outlet for stock than the Victoria River, which was in favour at the time.

The cattle were on the road again in March, after a hold up in Katherine awaiting supplies from Darwin. Tom and Paddy Cahill with Hugh and Wattie Gordon went ahead with the Wave Hill Mob. The nature of the countryside changed to sparsely timbered undulating blacksoil plains and downs and the whole mob was clearly visible, which made the job for horses and stockmen much easier. Ahead, between Delamere Station and the Negri River they encountered basalt country and the ground was rocky, so the horses had to be shod.

At O'D Creek the Ord and Wave Hill cattle went their separate ways. The Ord cattle had about 160 miles of virgin country to cross before arriving at their destination in Western Australia. They roughly followed Gregory's route and after crossing the Victoria River, Nat piloted the mob to the Wickham River and then through a spectacular gorge. The 2,500 head of cattle and horses had to travel for miles through the narrow, rugged passage, with a steep rock wall on one side and a deep waterhole on the other. After many hours of slow travelling the stock and men emerged from the jagged red and brown gorge dotted with fig trees, cabbage and feather palms. Everyone was relieved to be out of the gorge because they would have been sitting ducks if the Aborigines had decided to attack.

Now the Wickham changed course so the mob was taken through rough country and put onto Depot Creek - so named because Gregory had a depot there. The country then opened out into the magnificent Roe Downs, ten to sixty miles wide. At Poison Creek seven or eight cattle died as a result of eating fuscia bush, but this was the only mishap. The trek continued through level spinifex country to Black Gin Creek and then onto Gill Creek before encountering a dry stretch to G.B. Creek. Being some distance from the coast this country enjoys a dry climate and experiences extremes of heat and cold.

Nat directed the cattle on a south westerly course to Stirling Creek which was followed until limestone gorges blocked their passage. Finding an alternative route up Murray Creek, he then piloted them to Campbell Spring, within forty miles of the Ord. Some slow travelling followed through very rough basalt ridges east of the Negri before they finally reached their destination on the 22nd June. The drive from Bogantungan to Ord River station had taken eighteen months!

As a result of some winter rain shortly before their arrival, fresh green grass greeted the travel-weary cattle. With his contract successfully completed, Nat turned the stock over to Bob Button. Donald Swan, G.W. Campbell, Paddy MacDonald, Octolonius Turtle Sinclair and Andy Giffney accepted the offer of three pounds a week to remain on and help Button set up the station. The remaining stockmen returned to the Territory.[12] After only a few days at the Ord the cook, Andy Giffney died of scurvy, a sad end to a new beginning.

Ord River Station remains on what was to be a temporary site, on a bend of Forrest Creek, two miles from its junction with the Ord River. Wells had to be dug as Forrest Creek had no permanent supply of water. Osmand inappropriately named the station Plympton St. Mary, but the name never stuck and it remained 'The Ord'.

Nat and Gordie went on a scouting trip up river to inspect the country praised so highly by W.J. O'Donnell and Alexander Forrest. O'Donnell with W. Carr-Boyd had made a journey of exploration from the Overland Telegraph line to the Kimberley region in 1883, and had described the country he saw in what was thought to be exaggerated terms. After camping one night on a long waterhole in a beautiful valley, forty miles above the station and beyond the junction of the Nicholson River, Nat and Gordie were convinced that O'Donnell's praise was in no way exaggerated.

The pair returned across plains and tree-studded downs via the Nicholson and Linacre Rivers. Gordie was riding ahead of Nat along the right bank of the Linacre when wisps of smoke were seen in the sky. Nat signalled silently to continue their course. Around a bend they suddenly came upon an Aboriginal camp and the startled occupants cleared out as fast as they could. The camp was on a spit of brush covered sand, a low narrow peninsula formed by the junction of the river and the Brook, a stream coming in from the right.

> *Only one was carrying spears, so most of the men must have been out hunting. Riding slowly into the camp we noted the evidence left by the silent stampede. There were some coolamons, a stand of spears, campfires, and bush and plaited grass windbreaks at each family semi-circle. We left everything untouched and rode on without seeing any more sign of them.*

In later years Gordie recalled that members of this same tribe were credited with the rescue of a hopelessly lost and perishing 'swaggie'. The 'swaggie' was leaving the Kimberley via the Ord in 1886 when he wandered from the ill-defined 'overland track', got hopelessly lost, was desperate for water and forced to abandon his swag. Other parties of Aborigines had run away from him, thinking him mad, as he tried by signs to explain his plight. He soon would have been, but for the aid of the less timid Aborigines who gave him water from a coolamon, and took him back to his swag by following his tracks. After feeding him for a day they put him on the road, indicating by gestures the general direction of the Ord River Station, where he eventually arrived safely.

Back at 'The Ord', Bob Button and Don Swan set off for Darwin with twenty packhorses to get supplies for the fledgling station.

Soon after Button and Swan returned from their mission to Darwin, they were surprised by the arrival at the station of Joe McMaugh. Joe had sailed from Darwin with 'Smelter' Bill Osmand on his yacht the Cushie Doo, anchoring in the Cambridge Gulf on the 20th August 1884. The main object of Osmand's exploration of the Cambridge Gulf area was to report on its suitability as a port. At that time the Northern Territory of South Australia was considering developing the Victoria River as a port to supply the Victoria Downs and East Kimberley area. Despite Osmand's unfavourable report on its prospects as a port the discovery of gold in the Kimberley soon placed the Cambridge Gulf on the map and the Victoria River option never eventuated.[13]

Osmand carried supplies for his new station and so he had despatched McMaugh, Sandy Mauger and two blackboys to find a road to the property and to inform Button that rations and equipment were waiting. With McMaugh as guide and a string of packhorses, Button made the first of many trips to Cambridge Gulf as manager of 'The Ord'.[14] As it turned out the mountain of supplies was more than the packhorses could carry so the excess was sealed in a container and buried to be collected later. Unfortunately the Aborigines found this cache and in their efforts to open the container lit a fire under it and ruined all the goods.

Following their scouting trip Nat and Gordie proceeded overland to Katherine and on the way met G.R. Hedley with a mob of 600 cattle. Commissioned by Nat back at Burketown, Hedley had followed on behind the Ord cattle to bring up any strays, weak or abandoned stock, lost en route.

After selling their horses at Burrundie, Nat and Gordie took a coach to Southport where a small steamer ferried them across the bay to Palmerston.[15] Gordie sailed for Sydney with Wattie Gordon on the S.S. *Naples* on the 16th October 1884. Nat had some business to attend to in Palmerston and sailed ten days later on the S.S. *Catterthun* in the company of Joe McMaugh and George Hedley.[16] Back in the 1940's Gordie commented on his father's third great droving expedition quite simply:

> *Here again this unerring bushman, aided somewhat by Bob Button,*
> *discovered the best stockroute, one which in most of its length is used*
> *by all traffic to this day.*

In the 1990's the majority of these early stockroutes have been made redundant by the use of motor transport for trucking stock to market.

References

1. G. Buchanan, in the ms. of *Old Bluey* gives the date of arrival at Ord River station as the 22/6/1884 and in *Packhorse and Waterhole* as the 15/6/1884. Swan in an interview for the *Western Mail*, 30/12/37 gives the date as the end of June.
2. *The Western Mail* 30/12/37.
3. *Australian Dictionary of Biography Vol. 5*. 1966. Melbourne University Press - re J.A. Panton.
4. *The Western Mail* 30/12/37.
5. Dorothy Jounquay. 1974. *The Isisford Story*. Brisbane.
6. Mary Durack. 1988. *Kings in Grass Castles*. Corgi.
7. W. Linklater & Lynda Tapp. 1968. *Gather No Moss*. Macmillan.
8. Alfred Giles (nd). *The First Settlement in the N.T.* Unpub ms. S.A. Archives A5837.
9. C.H. Eden. 1872. *My Wife and I in Queensland*. London.
10. Edwina M. Edkins. (nd) *Reminiscences of Edwina M. Edkins*. Unpublished ms. held at Bimbah, Longreach Qld.
11. *Northern Territory Times and Gaette*, 29/12/1883
12. *The Western Mail*, 30/12/37.
13. *Northern Territory Times and Gazette*, 18/10/1884.
14. *Northern Territory Times and Gazette*, 18/10/1884.
15. Southport was a flourishing settlement at the mouth of the Blackmore River on the southern extremity of Darwin Harbour. In the early days goods were landed here because it was closer to the Pine Creek goldfields and saved many miles of cartage. Southport eventually died when it was bypassed by the railway line, which opened in 1889. Palmerston was the original name given to what is now Darwin.
16. Port Darwin Shipping and Passenger lists, NT Archives.

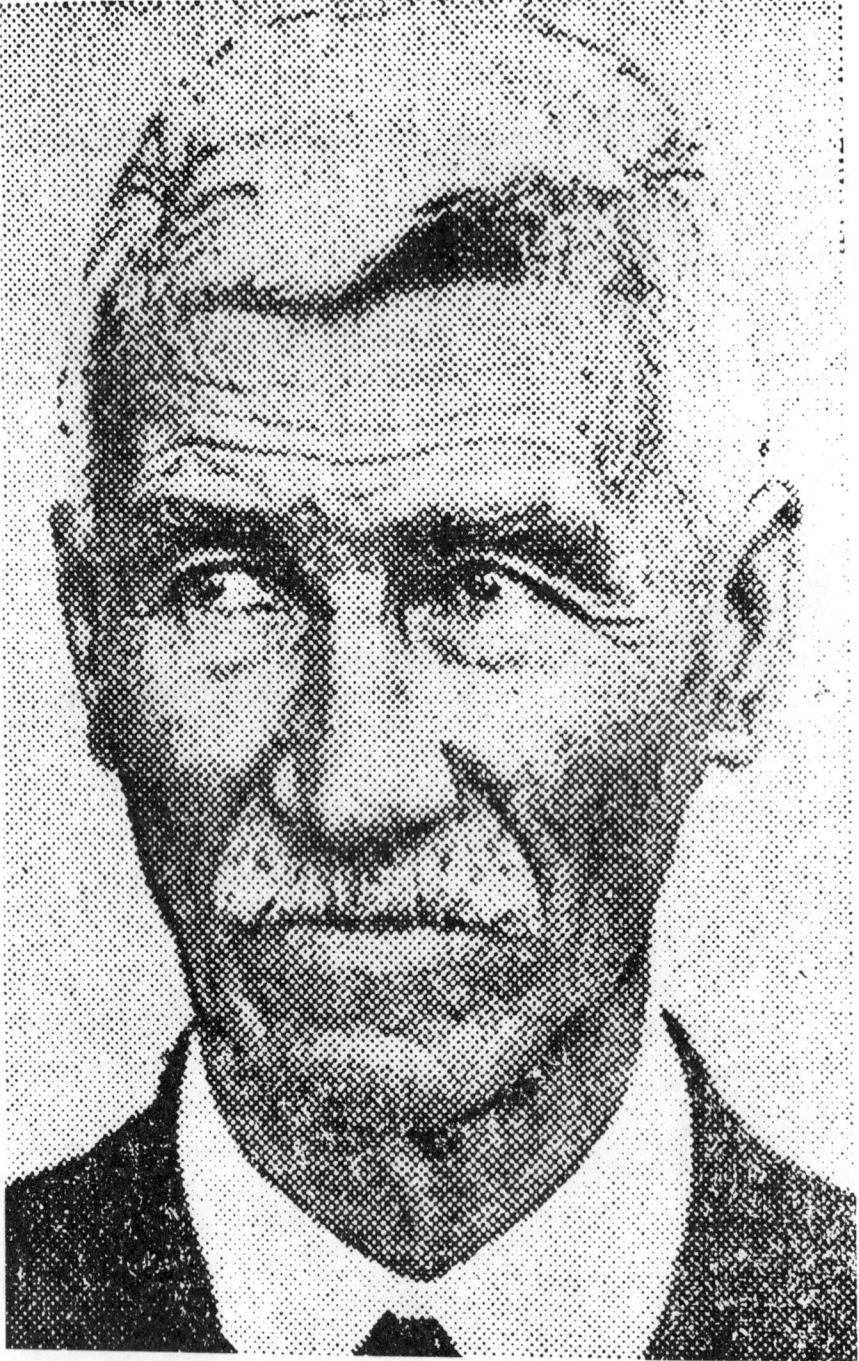

18. Donald Swan - A stockman with the first cattle delivered to Ord River Station
The Western Mail, 30/12/37

16. Gordon (Gordie) Buchanan, Nat's son who joined the Ord River drive
Buchanan Collection

17. Robert (Bob) Button, aged 28 years (Circa 1882) - First manager of Ord River Station
Courtesy of Battye Library, 67851P

12.

Wave Hill Genesis

1883 - 1885

Wave Hill Station represented Nat's highest hopes and greatest disappointment. The property he selected in the Victoria River district of the Northern Territory was the kind of country he most loved - open, tree-studded plains, well-grassed and watered. Here he felt he could finally cease his wandering and establish the long-awaited home for his wife and himself. For ten years he worked tirelessly to fulfil this ambition, but it was not to be.

Nat first became aware of the Victoria River district in 1859 through the Commissioner for Lands and Surveyor General in Queensland, Augustus Gregory. In 1855-6 Gregory explored the Victoria River area of the Northern Territory and traced Sturt Creek for some three hundred miles. Upon his return he wrote a very positive report on the country. In January 1878, Nat acted on Gregory's recommendation and without first inspecting the country, made pastoral lease applications.[1]

In December 1879, following his first trip across the Gulf Track and an exploratory trip with Sam Croker, Nat abandoned the 1878 leases and applied for eight blocks further south which followed the Victoria River.[2] In December 1881, he transferred some blocks to Fisher & Lyons but retained four which became the nucleus of Wave Hill.[3] Gregory's description of the country as the finest pastoral lands in Australia and equal to the best parts of Queensland for fattening purposes proved to be true. The landscape was of open plains and bald hills, mainly basaltic with some limestone, growing an abundance of Mitchell and blue grass, pigweed, wild melons and other herbage. It was well-watered back from the Victoria River and its tributaries, and timber for buildings, yards and fences was in plentiful supply. The two main drawbacks were the stony nature of the country, making horseshoeing a necessity and the weather - hot and dry pre-wet season, then so wet it made travel impossible.

Hugh and Wattie Gordon were Nat's partners in Wave Hill and they contributed financially to the stocking and establishment costs as well as to the management of the property. The first stock arrived in May 1883, with the delivery of 500 heifers from Donald McIntyre's Dalgonally Station in Queensland.

In October 1882 when Nat was supervising the movement of 20,000 head of Fisher & Lyons cattle to Glencoe, the first draft of stock for Wave Hill was making its way from Queensland. These cattle were brought across to the Overland Telegraph line via the Gulf Track, in charge of a drover named W.L. Laing. Wattie Gordon joined the drive some time after its commencement and saw it through to its destination. Laing was an unfortunate choice as drover because he proved to be a lazy bully who caused a lot of bad feeling in the camp. Once away from civilisation he displayed his true colours, and took advantage of Wattie's kindly nature and financial interest in the cattle. The men hated him and only their loyalty and high regard for Wattie stopped them from deserting. Laing obviously knew he was hated because night and day he wore a Derringer pistol strapped to his wrist. Once the party reached the Roper, the going was easier so a number of men pulled out. Only Laing, Gordon and the cook remained with the cattle. Sixty miles from the Elsey the cook became too ill to travel, so Laing set out to get an extra hand and more supplies.[4]

As the provisions were immediately obtainable at the Elsey, Laing should have returned to camp in four days, instead he took three long weeks. During this period Wattie held the cattle around the clock, and mustered the horses every morning. The invalid cook who could barely walk, provided rough meals and enabled Wattie to get some broken sleep by calling him when the cattle became restless at night. They were very lucky there wasn't a rush or trouble with Aborigines. Just beyond the Elsey, Hugh Gordon and Sam Croker rode to meet the cattle and Laing was promptly discharged. Piloted by Hugh Gordon, a much happier party passed up the King River near Katherine at the end of March, and reached the upper Victoria without further trouble.[5]

The first camp on Wave Hill was on the bank of a long deep waterhole on the Victoria River near its junction with Wattie Creek.[6] Here the first cattle on the Victoria River were released. Sam, Hugh and Wattie constructed bush timber huts and erected a horse paddock, the only fence on the property. The name Wave Hill was suggested by Sam who pointed out the sharp, wave-like undulations of the range on the eastern side of the Victoria River Valley.

One afternoon, shortly after their arrival, the local Aborigines descended upon the camp. The dogs barked a warning and the large group of raiders were surprised in the act of stealing provisions and vital equipment, such as buckets and billies, axes and shovels. Sam Croker was in charge and he only knew one way to handle the situation - shoot first and ask questions later. One of the Aborigines was shot dead by Sam as he attempted to escape across the river and the men were mystified to discover the dead man, of approximately thirty years of age, to be of mixed blood. Conjecture was rife as to whether he was a descendent of a member of either Leichhardt's or Gregory's exploring parties. Following this initial clash, perhaps partly as a consequence of it, open hostility reigned and Wave Hill became a hot spot to hold in those early years.

When first established, Wave Hill was extremely isolated with the nearest occupied property, Delamere, on the Flora River, approximately 120 miles to the north east. Sam Croker and Surveyor A.J. Woods had explored that area in 1880, prior to it being stocked with sheep for the owner Dr. W.J. Browne, in mid-1881.[7]

Fisher & Lyon's Victoria Downs Station lay eighty miles to the north of Wave Hill. It was formed in September 1883, and stocked with cattle delivered by Willie Gordon.[8]

Gordon later shifted 6,000 head from Glencoe to Victoria Downs.

Lindsay Crawford was appointed first manager. Nat had commended him to C.B. Fisher, when as manager of Richmond Downs, Crawford had impressed him with his competence.

> We understand that the recent attempt to form a depot on the Victoria River by Messrs. Fisher & Lyons for the purpose of supplying their cattle stations with stores, has at last, been successful, and the rations are now on the run. A good road from the depot, 60 miles distant from the headstation has been found and the teams have made the first trip. Owing to the number of stations in formation in this part of the country, amongst which are Osmand and Pantons, Buchanan Bros., and Fisher & Lyons, this matter is one that will have a very important bearing on the future of the pastoral lands of the Ord and Victoria Rivers, as by opening up this route, the difficulties and expenses of carting overland (some six hundred miles) have been overcome.
>
> Northern Territory Times & Gazette, 27/9/1884.

The nearest mail and telegraph station was at Katherine 300 miles away and initially all provisions and station supplies were packed overland from there. The trip took twelve days and the cost of delivery was 250 pounds per ton.[9] This financial burden was lightened in 1884, when Victoria Downs station established the Depot Landing on the Victoria River. Supplies from Darwin were then shipped up the Victoria to the Depot. Here they were collected by teamsters and carried via Jasper Gorge to the station, at twenty pounds per ton - quite a saving!

Travellers anticipated their passage through the Gorge with trepidation. The path through the red rock walls was narrow and provided plenty of cover, which made it the ideal place for the Aborigines to stage an ambush. Attacks became so frequent and severe that eventually a police escort was offered to travellers passing through the gorge.[10]

In May 1884, coinciding with the arrival of 860 Dalgonally cattle, delivered by Tom Cahill, the Wave Hill pastoral lease applications were converted to twenty year pastoral leases. Twenty five bulls, 1,000 heifers and some horses started the journey to Wave Hill, but the outbreak of pleuro accounted for the loss of 165 of the cattle.[11] The stocking terms in the Territory were relatively easy, being ten head of small cattle or two head of large cattle per square mile. For the first seven years of the twenty five year lease the cost was sixpence per square mile, and two shillings and sixpence per square mile thereafter. By the end of 1885 there were between 4,000 and 5,000 mixed cattle on the station.[12]

In 1883, Nat's attorney in Sydney, William Kilgour, arranged a partnership between Nat and his wealthy brother, W.F. Buchanan. W.F. had taken up pastoral leases on Sturt Creek, adjoining Wave Hill, in 1881.[13] Gordie was of the opinion that this prompted Kilgour to set up the partnership, giving W.F. a half share. Kilgour had power-of-attorney in Nat's absence and the partnership was set up without his instructions or knowledge. Considering that Nat had already formed a working partnership with Hugh and Wattie Gordon - possibly in word only - it is unlikely he would have agreed to Kilgour's arrangement. Kilgour probably acted in good faith, believing the partnership would be of mutual benefit to the brothers. It would be interesting to know who initiated the proposal.

William Frederick Buchanan had built up a huge pastoral empire with property in New South Wales and Queensland, and specialised in breeding high quality stock. A strict authoritarian, he was hard working and an expert horseman and stockman and a shrewd businessman who displayed foresight and common sense. Although charitable when it suited his purpose, W.F. was considered to be very 'canny' with his money and appears to have shown no prior interest in becoming involved in Nat's affairs.[14] It may be that his desire to gain a foothold in the Victoria River District was initiated by the fact that other prominent businessmen of the day, such as Fisher & Lyons, were landholders in the area.

At the end of 1884, Nat and the Gordon brothers bought 3,000 East Darr cattle at Cloncurry in Queensland. The purchase price was over two pounds per head and necessitated a hefty overdraft at eight per cent interest. Hugh Gordon and Tom Cahill were responsible for the delivery of the two mobs to Wave Hill.

Disaster struck the mob in the form of redwater fever which struck down 1,200 of the cattle in a fortnight. The losses occurred in October 1885, between Abraham's Billabong and the King River, and were the first recorded big losses of travelling cattle from tick fever in the Northern Territory. A few months earlier several mobs of cattle bound for East Kimberley had passed the same spot without infection. Redwater had only appeared once before, and this was in 1882 on Glencoe among newly arrived stock from Queensland. When signs of illness developed in his mob Hugh Gordon suspected that some poisonous weed was the cause and hurried the cattle on, unknowingly adding to the losses. The depleted mob was finally delivered to Wave Hill towards the end of 1885. This ended the stocking phase.

The country lived up to its reputation for cattle breeding and fattening and the herds multiplied under the administration of Sam Croker and his experienced offsiders, Tom Cahill and the Gordon brothers.

References

1. Plan showing approximate positions of Pastoral Claims in the Northern Territory of South Australia, January 1878.
2. The 1883 pastoral claims map of the area shows W.F. Buchanan as lessee, but Nathaniel Buchanan is shown as the lessee in the Pastoral Lease Register. Pastoral Lease Numbers 853, 854, 855 and 856 were Nat Buchanan's leases.
3. *Northern Territory Pastoral Application Book 1*, N.T.A. F670.
4. *Northern Territory Times & Gazette*, 17/2/1883
5. *Northern Territory Times & Gazette*, 16/2/1884
6. *Hoofs and Horns*, June 1949.
7. *Government Resident's Report of the Northern Territory of South Australia*, 1/6/1884. NARU DU392.
8. *The Sydney Morning Herald* 24/2/22; *Northern Territory Times & Gazette*, 14/7/1883
9. Alfred Giles. (nd). *The First Pastoral Settlement in the N.T.* Unpub. ms. S.A. Archives A5837.
10. Police escorts for travellers through Jasper Gorge commenced in 1894.
11. *Northern Territory Times & Gazette,* 16/2/1884.
12. *Government Resident's Report of the Northern Territory and South Australia,* 31/12/1885. NARU DU392.
13. Telegram from the Surveyor General to the Government Resident dated 9/5/1881 re W.F. Buchanan's application for pastoral leases in the Sturt Creek area. N.T.R.S. A4644.
14. *The Australian Dictionary of Biography Vol. 3*. 1966. Melbourne University Press. Ref. W.F. Buchanan.

19. Wave-like formations on the hill which inspired Sam Croker to name the property Wave Hill
Photo courtesy of Darrell Lewis

20. Wave Hill Homestead - drawing by Charlie Flannagan
South Australian Museum, Album titled "Drawings by an Aboriginal" No. 24.

21. Heavily laden drays at the Depot Landing, VRD
Northern Territory Library, Historic Photos, PH 275/49

22. Vehicle passing through Jasper Gorge (Circa 1930)
Buchanan Collection

13.

Blazing the Murranji Track

1885 - 1886

—❧❧❧❦❧❧❧—

Wild dogs howl and the hedgewood groans,
A night wind whistles in semitones,
And bower birds play with human bones
Under a vacant sky.
The drover's mob is a cloud of dust,
The drover's mob is a sacred trust
Where the Devil says 'Can't!' and God says 'Must!'
Out on the Murranji.

Hoofs & Horns, June 1949.

Wave Hill was excellent country for horses and following the introduction of some well bred stallions and mares, quite a number of good gallopers were bred on the run. With all the new stations being set up in the Territory, horses were in demand so Nat decided to overland two mobs from New England. The first mob in 1885 was for sale and the second in 1886 for Wave Hill. On the second trip he blazed the famous Murranji Track.

In January 1885, after spending Christmas with Catherine in Walcha, Nat and Gordie, with the assistance of Jack Furnifull, bought 130 saddle horses from

Tamworth and Walcha. At Armidale they were joined by James Mitchell with forty horses of his own. Members of the party were Nat, Gordie, Jack Furnifull, James Mitchell, Willie Glass and a man named Craigie.

Between Bolivia and Warwick the stock route was bare and the adjoining paddocks were very inviting. The opportunity for a few hours free agistment was too tempting to resist. The next time they risked it, just beyond Roma, they had a narrow escape. Around midnight, after having helped themselves to five hours forbidden agistment, they were caught by a station hand in the act of hurrying the horses back onto the stockroute. Jack Furnifull, whose personality, appearance and gift of the gab enabled him to assume a simple and innocent attitude, managed to appease the indignant station employee. Jack was good at talking his way out of sticky situations but he didn't do quite as well the next time:

> *Taking a short cut from Augathella to Tambo through unfenced country heavily timbered with brigalow and mulga, we came upon a cattle mustering camp. We pulled up some distance away and the head stockman from the neighbouring camp came over to investigate. He hailed Furnifull, 'Where the hell do you think you're going? You are trespassing, you'll have to pay for this!' He then noticed my father and added, 'Who is that old bloke over there?' pointing in father's direction, 'I believe it's Old Bluey Buchanan! No one else would try that short cut through the scrub. Anyway it will cost you a fiver.' Jack, who was an inveterate bargainer, got it down to two pounds, but the head stockman wouldn't give him a receipt. 'Give it to the Tambo Hospital and we'll cry quits,' said Jack, but the hospital did not benefit by that transaction.*

The travelling horses averaged about eighteen miles a day, and at every town the travelling stock permit was submitted to the police, who were supposed to check the brands. Only in a few towns where Nat's name was unknown were the brands on the horses given even a cursory inspection. The brands on the 170 horses were properly checked on one occasion and then only because there was a new policeman in town. Many of the brands on the herd were indistinct so it took him three hours and probably curbed his devotion to the rule book in the future.

At Blackall the horses were put in the police yard for the night. Someone inadvertently left the back gate open and around dusk the horses escaped into a back street, scattering fowls, goats and children in all directions. Gathering pace, they thundered into the main street, filling the whole roadway and making it impossible for a horseman to get to the leaders. They galloped on, carrying away several panels of the fencing of a big paddock. The mob was successfully recovered next day and apart from a few minor cuts and bruises, the horses came to no serious harm.

At night only the working horses were hobbled. Because of their quarrelsome nature in a mob and their need to feed more frequently than cattle the horses were unrestrained and unsupervised at night and mustered each morning.

At a billabong near Aramac while the men stopped to unpack the 'tucker horse' for a quick lunch, the horses bolted. For no apparent reason one of the mob imagined he heard or saw something, and the rush was on. A snort, and all ears pricked and heads lifted excitedly. In a moment they were off, their hooves striking the ground like a roll of thunder and in their wake a long cloud of dust obscured the herd from view. Not wanting to miss the excitement one of the saddle horses broke away and joined the mob which galloped some miles to reach the open downs before the men could steady them. In their wake they left a trail of torn swags and camp gear which had to be collected and repacked.

The first parting of the ways came at Muttaburra where Jim Mitchell turned off to Darr River Downs, taking his forty horses with him. Then at Dalgonally Jack Furnifull and Craigie left to go cattle droving to Bourke, New South Wales. After a brief stop at Dalgonally, with Nat's old friend Donald McIntyre, the remaining three men went on with the mob of ninety head.

They made good time along the Gulf Road and caught up with Hugh, Wattie and Tom Cahill with 3,000 cattle from Queensland for Wave Hill. At McMinn's Bluff on the Roper, they overtook D'Arcy Uhr and two of the Morck brothers with a mob of cattle for Newcastle Waters. The rush to stock Territory leases was continuing.

On arrival in Katherine the horses were promptly sold to R.S. McPhee, who took them on to the Kimberley goldfields. Nat and Gordie did a quick turn around and on the 22nd July, 1885, boarded the S.S. *Tannadice* in Darwin and returned to Sydney.[1]

By Christmas 1885, Nat and Gordie were in Gayndah, a few miles from Gordie's birthplace at Ban Ban Station. After celebrating the festive season with their friends they began preparing for the second overland trip with horses. They bought a few at Yenda and Taroom and rode on to Roma where Jack Furnifull and Willie Glass had collected a mob of 100 to take on to Wave Hill.

> *The day we left Roma the post office thermometer read 118 degrees Fahrenheit in the shade. We travelled by the usual route, visiting old friends of my father's on the way. At Dalgonally we were warmly welcomed by Donald McIntyre, and as was Bluey's custom we spelled there for a week. It was here, once again, that Jack Furnifull left us to lift another mob of fats to Bourke, and we welcomed the addition of Archie Ferguson and Mick Barry to our little party. These two men were to stay on at Wave Hill as stockmen.*

Instead of taking the Gulf Track, Nat took the inland route from Turn off Lagoon on the Nicholson River, across the Barkly. The Tablelands were now sparsely settled and in good seasons when surface water was available this route was used by inland pastoralists. Perhaps Nat was already toying with the idea of finding a more direct route to Wave Hill. After following up the Nicholson River, they mounted the coast range and struck the head of Cresswell Creek. The party camped the night at Cresswell Downs and the owner, Tom Perry, accompanied them on the next day's stage to Anthony Lagoon. There was a carrier with his wife and family camped there and they invited the travellers to tea.

> *That year on the Barkly Tablelands there was a rat plague, millions of the little rodents, and the blacks waxed fat on them. Tom Perry suggested that the hospitable carrier might give us rats for tea, and we were joking about this possibility when one of his boys came running down to our camp with the message, 'Mum says to come up to tea, the rats is gettin' cold!' Tom, who knew my father's fastidious appetite well, apologised profusely.*

> *Nat abominated pig and would not even eat a bandicoot unless hard pressed by hunger, because he thought it tasted like pork. I remember once when we were staying with my mother and grandmother in a hotel in Goulburn, New South Wales, where they had a private dining and sitting room, a waitress brought in a covered tray and set it before him. When he lifted the cover there in resplendently decked nakedness, was a sucking pig. Quickly replacing the cover with a look of disgust, and not asking the others if they would care for any, he said to the astonished waitress in a quiet and pained tone of voice, 'Take it away. SOMEONE might like it!'*

At Anthony Lagoon, over 400 miles east of Wave Hill, they were interested to hear news that Sam Croker had recently passed through collecting strayed cattle. It was common for cattle and horses to drift back to where they were bred so when they were scared by Aborigines, their instincts lead them eastwards. Cattle also tend to graze against the wind, which in the Territory and East Kimberley blows from the east and south-east during the dry season.

Between Eva Downs and Powell Creek was about 80 miles, which included two dry stages. At Monmoona Creek, the only water on this stretch, they were fortunate to meet a friendly Aborigine who led them to the only waterhole. In the morning they reached Powell Creek and were greeted by a surprised Sam Croker who was heading home after collecting ninety head of Wave Hill strays. The two parties joined forces and passing Newcastle Waters continued north of Sturt's Plain to Frew's Ponds, where a united camp was set up. Here they were disturbed by the mystery deaths of two horses, for which they blamed poison weed.

Between their camp and the head of the Armstrong River to the west, lay well over 100 trackless, possibly waterless miles. A few months earlier, Croker with two Aborigines had crossed this way tracking the strays but had taken a more southerly route, avoiding the hedgewood scrub. They didn't discover any waterholes and made the crossing with the aid of a few well-timed showers of rain. Even earlier, in June 1885, well known bushman and drover, G. R. Hedley, crossed from Newcastle Waters to Victoria Downs, on a track to the north of Sam's route. With no Aboriginal guides he missed the waterholes and for seventy waterless miles battled through the bulwaddy scrubs.[2] Neither Hedley nor Croker were slowed down by travelling stock and the necessity to find large quantities of surface water when they made their crossings.

Finding a stockroute across this daunting country would benefit all Victoria River district cattlemen by providing a much shorter route to Queensland markets, and there was also the possibility of opening up the Kimberley goldfields market to Barkly stock. Nat and Sam accepted the challenge despite the considerable risk of losing their horses and cattle if they failed. This was not a journey to be undertaken by the faint-hearted.

In the early 1860's, the great explorer John McDouall Stuart had made determined efforts to get through this way to the Victoria River. From several points between Frew's Ponds and Newcastle Waters, he tried to penetrate that vast, waterless forest of hedgewood and lancewood scrub without success. He made several abortive attempts to cross, some of his horses died of thirst, and a lot of time was lost before he gave up and went north instead. Stuart made the following entry in his journal:

> *The scrub we were compelled to return from was the thickest I have ever had to contend with. The horses would not face it. They turned about in every direction, and we were in danger of losing them. In two or three yards they were quite out of sight. In the short distance we penetrated it has torn our 2 hands, faces, clothes, and, what is of more consequence, our saddle-bags, all to pieces. It consists of scrub of every kind, which is as thick as a hedge.*

Before setting out Nat decided to telegraph his intention of making the crossing, so with Gordie he rode north to Daly Waters telegraph station. This was reported in the *Northern Territory Gazette*, 28 August 1886.

In those heady days when the Kimberley goldrush was at its height the inland tracks were alive with travellers from other colonies. When returning to Frew's Ponds, Nat and Gordie encountered a party camped at Milner's Lagoon. They were goldseekers from Queensland and South Australia, heading for the Kimberley. Being of like mind and destination, they had banded together at Powell Creek to become travelling companions on the long road ahead via Katherine. They were a pretty rough looking crew, poorly dressed in a strange variety of worn out clothes, but they were high spirited and had good horses. Nat noticed they numbered thirteen - what he called the Devil's number - and commented on their ragged attire. From then on this band was known as the 'Ragged Thirteen'. In the Territory and across to Halls Creek the 'Ragged Thirteen' earned a reputation for pranks and lawlessness.[3]

When Nat casually mentioned to the group that he intended to strike straight through to the Victoria, they thought him mad. One of the group, Sandy Myrtle, explained to Nat that even a famous explorer like Stuart didn't make it. Nat replied, 'Well, you know, I'm a bit of an explorer myself, and so is Sam Croker. I think we can get hold of some blacks that live in that country to give us a start.'

'So long then, good luck, meet you in Kimberley,' replied Sandy.[4]

The Buchanans headed south to their camp at Frew's Ponds and the 'Ragged Thirteen' went north to Abraham's Billabong where their riotous behaviour disrupted the 'one horse' settlement for days. Mat Kirwan kept a store at Abrahams, and the day before the 'Thirteen' left there was a bullock hanging on the gallows. Broke, because of their drinking and gambling spree, they decided to steal half the carcase, but were caught in the act. Mat had a reputation in the Territory for fighting but he had been weakened by an attack of malaria. Disregarding this, he challenged the leader of the 'Thirteen' to bring out his best pugilist and they would fight for the right to the beef. Holmes, taking into consideration Mat's weakened condition decided to even up the odds by producing his second best fighter, Sandy Myrtle.[5] Poor old Mat was so wasted by fever that he only lasted two rounds and the gang rode off jubilantly with the prize.

The 'Thirteen' continued to pillage their way through the Territory eventually arriving at Gordon Creek, which was then the head station of Victoria Downs. The manager, who was out mustering with his men left instructions with the cook not to sell anything to travellers. The 'Ragged Thirteen' were out of flour, and when the cook refused to open the store or give over the key they simply stripped a sheet of galvanised iron from the end of the building and took what flour they wanted. They were a casual lot and realised their strength so did not hurry away. They carried salt and rifles, and there was plenty of beef on the hoof to be had along most of the route. The manager, Lindsay Crawford, had to accept their immunity 'beyond the reach and rule of law.'

When Nat arrived at the Frew's Pond camp he found that Sam had been busy in his absence. He had managed to contact some local Aborigines who agreed to lead the party to a big waterhole far to the west. Before setting out, Nat placed an upturned bucket on a stump by a small waterhole to mark the start of the route. The creek became known as Bucket Creek and the waterhole was referred to by drovers ever after as 'The Bucket'. The Aboriginal guides proved their reliability when after two days and one night of travelling, they led the thirsty stock and travel-weary men to Murranji Waterhole. It was a forbidding place surrounded by thirty foot high lancewood scrub interspersed with bulwaddy.[6] Murranji was a local Aboriginal word meaning frog.

After a short rest and the assurance of their guides that there was 'big fella water' further on, the party moved forward in the direction of Yellow Waterholes. Their main worry was whether the water referred to by the Aborigines would be sufficient for 190 head of cattle and horses.

> The Yellow Waterholes did not fail us. The country we had passed over was flat to undulating, and covered by dense thickets of lancewood and bulwaddy scrub which made travelling with stock very trying. By travelling night and day both dry stages were safely passed, but on the second stage short delays occurred when our guides unwittingly led us into cul-de-sacs of bulwaddy scrub. Long glades or avenues in this impenetrable forest closed up at the further end and forced returns and detours.

Bulwaddy extends its hardwood branches outwards from the ground interlocking with its neighbours to form an impenetrable barrier to stock and riders. Anthills which studded the area contributed to the hazards faced by the horsemen.

> At Yellow Waterholes another two horses died, probably the result of eating a noxious weed. Before moving on we killed a beast, part of which we gave to our black entourage, now increased by local natives, reassured by our apparent peaceful motives and attracted by the prospect of sharing in the distribution of beef and tobacco.[7]

The Murranji Track terminated at Top Springs on a tributary of the Victoria River called the Armstrong. From there the track branched in three directions, Katherine, Victoria Downs and Wave Hill. The path to the latter was via Armstrong River to the Victoria River.

> We were the first white men to visit the two waters, Murranji and Yellow Waterholes, and the first accompanied by stock to cross this way. Thus was opened the greatest of short cuts ever discovered in the north. For overlanders from Queensland and South Australia en route to the Victoria River District, the Kimberley goldfields and Wyndham, this track eliminated about three hundred miles of travel. In later years it was to become a highway for thousands of cattle travelling from west to east to link up with the stockroute across the Barkly Tablelands that roughly followed the track taken by Old Bluey and Sam Croker in their 1877 exploration.

Initially the Murranji track was mainly used by goldseekers taking a short cut to the Kimberley. The track gained a formidable reputation because of attacks by Aborigines and deaths of travellers from thirst. The development of the Murranji as a regular stockroute was delayed for several reasons. Firstly, by the time the Murranji was successfully penetrated by stock and packhorses the stocking of the western stations had already taken place. Secondly, the appearance and increase of redwater fever soon led Queensland and Western Australia to close their borders against the introduction of Territory cattle. Thirdly both Murranji and Yellow Waterholes could not provide year-round surface water. It appears that the track was infrequently used by drovers until 1904, when a combination of the effects of redwater fever and drought decimated Queensland cattle, making restocking necessary. Up to 10,000 cattle in six separate mobs made the crossing from west to east across the Murranji in that year.[8]

Before the Government sunk bores along the track in 1924, the Murranji drovers led a hard and dangerous life. They had to contend with long dry stages through almost impenetrable scrub, either dusty or boggy conditions depending on the season, noxious plants which poisoned the cattle, and hostile Aborigines. A unique hazard was the 'drummy' nature of the limestone ground which spooked cattle, causing bad rushes. These rushes were particularly difficult for the drovers to control because of the confinement of the dense and dangerous scrub which caused injuries to men and loss of stock.

The droving record for the longest dry stage ever travelled by a large mob of cattle was set on the Murranji in 1909 by Nat's nephews, the Farquharson brothers of Inverway station. They started more than 1,000 bullocks on the track east and found both the Yellow and Murranji Waterholes dry. Blessed by cool, overcast conditions, the cattle were driven around the clock. After dark hurricane lamps were used to give them a lead. They crossed 110 waterless miles with negligible stock losses.[9]

Even after bores were sunk at regular intervals and a wide path cleared through the scrub, droving conditions were still difficult and the track retained its infamous reputation. The Murranji saw its heaviest traffic following the bombing of Darwin and the closure of Wyndham meatworks during the Second World War. Congestion on the track and failure of the basic watering facilities to cope with demand caused many headaches for drovers. In the early 1960's when trucking cattle superseded droving, the Murranji fell into disuse.[10]

Darrell Lewis, who wrote a report on the history of the Murranji for the National Trust, summed it up: 'It has truly become the ghost road of the drovers, but the legend of the Murranji and the Murranji drovers will always remain as one of the enduring symbols of Australia's outback heritage.'

Nov 49

References

1. Darwin Shipping and Passenger lists, NT Archives and T*he Northern Territory Times & Gazette*, 25/7/1885.
2. *The Northern Territory Times & Gazette*, 29/8/85.
3. The names of the individual members of the 'Ragged Thirteen' are recorded in *Gather No Moss*, W. Linklater and L. Tapp, 1968.
4. Sandy Myrtle once managed Myrtle station in S.A. and from this he took his assumed name. His real name was McDonald.
5. Tom Holmes was the leader of the 'Ragged Thirteen'. He later took up Banka Banka station under the name of Tom Nugent. C.E.Flinders. (nd). *Forty Five Years in the Great Nor-West of Western Australia*, unpub. ms. Jack Campbell was the cook and fighter of the group. C.E.Flinders, op. cit.
6. Lancewood is Acacia shirleyi, an erect bushy tree that grows in dense stands.
7. Ironwood (Erythopheleum chlorostachys) which is highly poisonous to all mammals is now thought to have been the cause of mysterious deaths of horses and cattle.
8. Darrell Lewis, 1992, *Ghost Road of the Drovers, A Report on the History and Historic Sites of the Murranji Stock Route*, Prepared for the National Trust of Australia (N.T.).
9. In an article by E. Hill in *Walkabout*, 1/11/49 the number of cattle lost is reported to be five.
10. D. Lewis, op cit.

THE MURRANJI TRACK

The Murranji Track

23. The Bucket Waterhole at the start of the Murranji Track
Photograph courtesy of Darrell Lewis

24. The Murranji Waterhole
Photograph courtesy of Darrell Lewis

25. The Yellow Waterhole
Photograph courtesy of Darrell Lewis

26. The Farquharsons of Inverway Station
From Left - Hugh, Harry, Archie
Photograph reproduced by Darrell Lewis from an unknown, undated
newspaper clipping in the Buchanan Collection

14.

The Stocking of Sturt Creek

1886

W.F. Buchanan took up leases in 1881 on Sturt Creek, about 120 miles to the south west of Wave Hill.[1] The Sturt Creek leases were conveniently closer to the Kimberley Goldfield's market than Wave Hill so it was decided to stock them.

A.C. Gregory had put Sturt Creek on the map in 1855-6 and it had remained untrodden by white men until 1878 when McDonald and Sullivan, using Gregory's route, followed the upper part of its course. Tracking the Victoria River to its head they struck south-west and came across Sturt Creek which they followed down for about sixty miles. They then rode east-south-east to Hookers Creek, and taking a north-easterly course hoped to reach Powell Creek on the Overland Telegraph line. Unable to find water within a few days, they were forced to backtrack to the Victoria River and returned to Katherine in August. McDonald and Sullivan found most of the country they explored disappointing, but they did say that the country on the Sturt was the best they had seen for sheep and cattle grazing.[2]

Alexander Forrest was the first to report the gold-bearing nature of the Kimberley country and his observations were confirmed by E.T. Hardman's expeditions to the area in 1884 and 1885. In August 1885, Hall and Slattery reported finding a gold nugget in the Black Elvire near what later became known as the town of Halls Creek. Wildly exaggerated reports soon reached all parts of the country. The rush was on. The Kimberley became the new Eldorado. Unprecedented numbers of prospectors and would-be prospectors poured in from all over Australia and New Zealand. The goldfields were officially proclaimed in May 1886 and by the end of that year there were about 3,000 men on the fields. By the end of 1888 it was reported that 10,000 men had come to the area, but of course many didn't stay long. The rush centred on Halls Creek which was situated about 200 miles south of Wyndham and 280 miles east of Derby. A large number of prospectors entered by these ports but others made the trek from the Overland Telegraph line, some travelling via Katherine and others chancing their luck on the uncertainties of the Murranji Track.[3]

Upon Nat and his party's arrival at Wave Hill following their successful Murranji crossing, they found the cattle to stock Sturt Creek were being mustered. With nothing much to do for a few days, Nat decided to do a little exploring. With Gordie and George, the Aboriginal tracker, he set out to investigate the headwaters of the Victoria River. They ran the river up for about fifty miles and found that it came from the west, as described by explorer Gregory. This trip would have taken him into the country that his nephews the Farquharson brothers later settled and named Inverway. By cutting the bends in the river they made a more direct trip back to the station and were delighted to discover some good open Mitchell grass country.

Back at Wave Hill they found they had been left behind. Sam Croker and Hugh Gordon had started 1,800 head of cattle for Sturt Creek in the second week of September.[4] The cattle were travelling in a westerly direction towards the Ord River Road, which they followed as far as Swan Creek. This route blazed by Nat in 1884 was now well defined by the traffic of gold seekers to the Kimberley. Nat, Gordie and George caught up with the cattle drive in a few days.

It was the driest time of the year and at that time nothing was known of the waters between the Swan and Sturt Creek. Their dilemma was whether they could find enough water at the nearest part of the Sturt to supply the big mob of cattle. The three latecomers with a packhorse carrying swags, some tucker and a water bag, set out for Sturt Creek from the top water on the Swan. Their mission was to find a sufficient supply of water for the cattle that would last until the wet season set in. Sam and Hugh were to follow in their tracks with the mob. The advance party ran up Swan Creek for about twelve miles in a southerly direction before leaving it to ride through rising but fairly level forest country until about midnight, when they tied the horses up for a spell. By four next morning they were once again in their saddles and after riding about four miles they emerged from the timber onto an open Mitchell grassed plain which stretched as far as the eye could see. Finding a shallow watercourse, they first followed it south-west, then south to a bend where it entered a wide plain. Here Gordie's horse, Sunbeam, pricked up his ears in interest and led them to a pearly lily-covered pool. The party stopped for breakfast by the pool and Gordie named it Sunbeam Hole. Later this site was chosen by Gordie's cousins, the Farquharson brothers, for the Inverway homestead.

The Farquharson brothers, Harry, Hughie and Archie, were of Scottish descent and natives of New England. They were the sons of Kate's sister Jessie who had married George Farquharson. The three took up the Inverway leases in about 1890, but they could not afford to stock them immediately, so they spent the next three years working as drovers, and buying and selling stock until they had the money for the nucleus of their herd. In 1894 they bought cattle in the Inverell district of northern New South Wales and drove them through Queensland to Inverway. It is not certain if they travelled via the Gulf Track or across the Barkly and onto the Murranji. The Aborigines on the station were so warlike at that time that one of the brothers always assumed the role of armed guard while the other two worked. Harry was the stockman, Hughie kept the books and Archie was jack of all trades. He planted a vegetable garden and built the homestead out of 2 feet wide, hand-adzed, upright bloodwood slabs. It was quite a few years before these battlers had any luxuries. In the days before motor transport goods were carried by camels to the property. Once when Harry was on a rare visit to Perth he decided to buy a wood stove but the problem of balancing the load on a camel was a difficult one. Harry solved it by buying two stoves.[5]

The brothers never married and led an isolated life as the following story from A.S. Bingle's manuscript, *This Is Our Country* illustrates.

> *Contact with the outside world from Inverway was difficult, as a somewhat gruesome story of that era shows. The Farquharsons' nearest neighbour was on Wave Hill station, 120 miles away. One day, a man who was optimistic enough to think he could cycle around Australia, arrived at Inverway. It was mid-afternoon and he was invited to have a cup of tea at a table in the shelter of the verandah.*
>
> *One brother pushed a cup of tea and the sugar toward the stranger and said, 'Here you are, this will help you on your way.' The traveller replied, 'I don't think so - I'll probably be dead in ten minutes.' Hughie, thinking the man had a quirkish sense of humour, remarked in his high-pitched, squeaky voice: 'Well, we've known you only five minutes, so you can't expect us to grieve.' But, as Hughie spoke, the man dropped something into his tea and rolled under the table. Literally, he was dead within ten minutes. The brothers looked at each other and one said drily, 'Well, old man, you were as good as your word'. Believing that in the tragic circumstances the man should not be buried before the police had been notified, the Farquharsons sent an Aboriginal footwalker to Wave Hill station, but that same night the wet season set in. The body was left under the table as the brothers thought it should not be moved before the police arrived. The weather was steamy and humid and it was not long before the smell of the swelling corpse was too pungent to bear, so they had to find another spot in which to eat their meals. Because of the storm delays it was ten days before the Aboriginal returned with the police and the body was*

duly buried, but it was many weeks afterwards before the Aboriginal
staff would go near the house as they were terrified of contacting the
spirit of the dead man.

The brothers only occasionally left the station and although living conditions
were rudimentary, they were most hospitable hosts. Harry died in Perth in 1943,
Hughie in Darwin in 1946 and Archie on Inverway in 1959 at the ripe old age of
eighty nine. A headstone marks his grave and also commemorates the other two
brothers who were buried elsewhere.

When Vesteys bought up many of the stations between the Victoria River
District and Halls Creek, at one stage Inverway was the only one they couldn't get
their hands on, so the station was nicknamed "Intheway". When the brothers died,
the property passed into the hands of the Leahy Brothers. In 1956 Pat Underwood
took it over. Inverway has since been divided into three stations, Riveren,
Inverway and Bunda, all of which remain in the very capable hands of members
of the Underwood family. The old homestead built by Archie is now gone but in
its place is a spacious outback home and the traditional Inverway hospitality still
prevails. Sunbeam Hole is close to the house and is bigger since Sunbeam Creek
has been partly dammed. It is a tranquil spot surrounded by beautiful trees. The
magnificent Buchanan yards, built by the Farquharsons near Mt. Farquharson and
adjacent to Buchanan Springs, although in an advanced state of decay, still
demonstrate the skill and endurance of these early pioneers.

Back in 1886 Sunbeam Hole could only supply enough water for a few days.
It was urgent that a larger source be found. By his dead reckoning Nat knew they
were near Sturt Creek. Across the plain to the south, about four miles distant was
a long line of trees which looked like coolibahs. This species of eucalypt often
lined the banks of creeks and billabongs of plains and downs country, and was a
good indication that there was a considerable watercourse ahead. The position,
size and trend of this creek corresponded with Gregory's map and description. It
had to be, and was, the 300 mile long Sturt Creek.

Upon reaching the Sturt they ran it up for three miles and to their great relief
found an ample supply of water in two well filled waterholes, both larger than
Sunbeam. One of the waterholes was half a mile long and appeared to be
permanent. A party of Aborigines was camped there but quickly scattered at the
approach of the strangers. Very probably these were the first white men they had
ever seen and so, not surprisingly, they were quite frightened. A few women and
piccaninnies climbed up coolibah trees as high as the branches would support
them, and Gordie was astonished when one dusky lady carried a dingo up the tree
with her. From the uncertain safety of the tree branches they continually and
loudly harangued the intruders. After stopping for a few hours rest the whites
moved on. George, used to the Wave Hill country, was excited by the lack of
stones in the area and announced to Nat, 'Good fella ground, can't miss 'em track'.

When they ran the creek down further they discovered two more large waterholes, the second being over half a mile long and eight feet deep. Sure now that they had discovered sufficient water for the cattle they turned back, and taking a different route up a tributary creek discovered a more open road for the travelling stock to follow. Sam and Hugh were too far away from Sunbeam Hole to reach it that day, so the horses were sent on to water and enough returned to provide mounts for the drovers the following day. The dry stage ended early next day without loss.

Four days later Croker formed a temporary camp on Emer Waterhole, about fifteen miles above Wallamunga - this was the first settlement on Sturt Creek. On the east side of the creek there was a low, thickly-timbered range with good grass and spinifex, while magnificent plains extended opposite. The horses were in poor condition following the Wave Hill calf muster and this trip, but responded miraculously to the fresh green pick along the Sturt.

Early in 1887 Stretch, Lewers, Weeks and Foster took up leases on the lower Sturt. They stocked the leases with 750 cattle and thirty horses from Normanton in Queensland, travelling via the Tablelands and arriving on Sturt Creek in October 1888. As a result of tick fever and pleuro 200 cattle died en route. This second settlement on Sturt Creek was named Denison Downs and after a great struggle the pioneer owners saw their small herd increase. They hung on until better times.[6]

Nat didn't stop still for long. Once the camp was established he set out with George and Gordie to explore Sturt Creek further down.

> *After travelling sixteen miles through beautiful country we came to a magnificent stretch of water, and a further twelve miles through the same country brought us to a reach of water 200 yards across and over a mile long, which we called Buchanan Lake. This was a very beautiful place. The lake, which was covered with wild fowl, was situated at the foot of a little ridge from where we viewed the surrounding luxuriant grassy plains extending to the horizon. To the north about eight miles away we saw an isolated peak rising from the top of high downs which we named Gregory Peak.*

They returned to Croker's camp on the 2nd October and stayed overnight before setting out next day in a north westerly direction for Ord River Station.[7] Three days later they arrived at Button's camp.

Nat and Gordie had a holiday with Kate and then returned to Port Darwin on the S.S. *Guthrie* in February. Next day when they sailed for Wyndham aboard the S.S. *Dickie* their fellow travellers included Patsy Durack, W. Carr-Boyd, and Willie Glass.[8] This 'cockleshell of a steamer' was overloaded with a riotous crowd of goldseekers who raided the cook's galley and took the ham which was reserved for the Captain's table. Gordie was very seasick on the thirty six hour journey, his distress somewhat relieved by champagne supplied by Patsy Durack.

Disembarking at Wyndham, Nat and Gordie joined the newly arrived prospectors and set out on horseback along the littered goldfields road to Halls Creek.

When the Territory market for fats declined, Nat decided to move the cattle from Sturt Creek to W.F. Buchanan's Elvire Block, twenty five miles from Halls Creek. This gave quicker access to the goldfields' market where a few could be sold to cover expenses. The trip started in June with Hugh Gordon in charge. The cattle had learned good road manners the year before and were all in peak condition. In fact, some were almost too fat for fast travel. At the end of the first day's stage they came to a pre-selected camping ground. Lowing from beasts well in front of the mob fell faintly on the drovers' ears and through a misty, golden, late afternoon dust haze they contentedly made their way onto the night's camp. As the tropic night fell Hugh Gordon sent his stockmen into camp and with the aid of the dog watch, rounded off the corners of the straggled herd. It took a little time to settle them. Cattle often form friendships and so had to find their mates and cows their calves. After the evening meal the first watch rode out to relieve Hugh who joined the other men in the glow of the camp fire.

Cattle on the road drink little in the morning so in preparation for the dry stage ahead they were grazed out on the downs until noon, then brought back to the waterhole for two hours to make certain that all had a full drink. Some waded up to their necks in water feeding on the giant water couch grass which grows in billabongs, river waterholes and creeks, after they cease flowing.

A crack of the whip was the signal to get together again and face the western sun. That night was a dry camp. With the aid of a three-quarter moon the mob was started before daylight the next morning. A slice of salt beef on a slab of damper, washed down with a pannikin of tea served as breakfast before the men set out on the next leg of the dry stage. The horses and plant arrived at water early and remounts were sent back. Late that night the cattle arrived and as there were no other stock in that part of the country, they were let loose.

The herd was easily mustered next morning and well-watered before facing a longer dry stage of about fifty miles between Wallamunga Waterhole and Button Creek. Open country and a good moon enabled them to travel at night but by the second morning they were still twenty miles from the Button Waterhole. The quicker-paced packhorse droving plant was already there so they sent back fresh horses, tucker and water.

The cattle were now too thirsty to eat, so to increase their pace they were extended in a mile long, narrow column in which they could walk at the rate of about three miles per hour. With one man in the lead to steer and to steady the pace, they trooped along, forming a bare track from which clouds of dust rose to drift through the open forest or across the downs. Speed was everything now but it must be organised and regulated to the slowest walking beast. One man rode ahead to control the leaders and another man rode up and down the plodding column to check the pace and to ensure orderly progress, and to keep in touch with the lead man. The main purpose was to prevent the cattle running and trotting, which would result in tongueing, distress, and possibly disaster at the waterhole.

It was afternoon before the coolibahs on the Button appeared three miles away across the open downs. A faint breeze from that direction soon caused a chorus of excited bellowing. The cattle had smelt water! Two men came up to steady them, leaving one with the ruck. Because the waterhole was not as full as expected the cattle were sent in to drink in lots of 100 or so while the remainder were held a half mile out to prevent them rushing in and spoiling the water. With no other water within reach, this would have meant the end of the road for most of the cattle. The stock were desperately thirsty and were only held with great difficulty. Two more stages of fifteen miles each were covered in three days before the mob was delivered to the Elvire in July. Nat took delivery of the mob and sent sixty three head into Halls Creek.[9] The remainder were put under the supervision of a Queenslander named Brock. Hugh Gordon and some of the men then returned to Wave Hill.

W.F. Buchanan's Elvire block became Nat's headquarters. From here he continued to pursue markets for Wave Hill fats until the property was sold in 1894. Sam Croker remained with Nat, to deliver fat bullocks to the various butchers on the goldfields. There were three butchers at intervals of fifteen, twenty five and forty miles from the Elvire. For a short time Nat had a butcher shop in a bough shed at Dry Creek where Sam Croker supplied part of the goldfields market with Sturt Creek and Wave Hill beef. [10]

The goldfields country was a mass of steep slate and quartz ridges, covered with buck spinifex and the few narrow grassy flats on the creeks were either rooted up by diggers or eaten out by their horses. For this reason the butchers could only hold two or three cattle at a time in their waterless and barren yards.

Gordie was instructed to bring in fifty head of killers from the Elvire lease and hold them on some good grassy country at Palm Springs, fifteen miles west of the town. Johnny Lyons, a butcher, came out periodically and bought a few as he wanted them, as did A.G. Layman and other butchers scattered about the fields. All the local graziers had cattle to sell to the butchers and competition was stiff, so it was a pretty cut-throat business.

In October a further 220 head arrived from Wave Hill and were held on the Elvire.[11] While the butchering business was progressing, it appears that Nat also dabbled in some prospecting, becoming involved with F. Burdett in the Evening Star claim. They must have had a falling out, because Burdett summoned Nat over working expenses for the mine, but the case was dismissed.[12] Further information about this episode is not forthcoming. By November, most of the 'killers' Gordie had been responsible for had been butchered and he left for Sydney in the company of Surveyor Nyulasy. They travelled overland to Derby where they boarded ship for Fremantle. Gordie then took a coach to Albany followed by a sea voyage to Sydney where he spent the whole of 1888, with Kate. The reason for this extended leave is not explained.

> *Buchanan has been slaughtering cattle from his Sturt's Creek Station, at the different diggings and selling prime beef at 6d. per lb. Sam Croker is managing the butchering. One thousand head of Wave Hill cattle are on the way from Wave Hill Station, Victoria River, to stock up Sturt's Creek. Splendid beef is being sold all over the diggings and at Wyndham,...*
>
> *Northern Territory Times and Gazette, 10/3/1888.*

The Darwin meat market continued to decline, putting more pressure on Nat to find fresh outlets for his cattle. In Halls Creek he continued to exploit the dwindling goldfields market with fats from Wave Hill and the Elvire. About eight months after delivering the cattle to the diggings a mob of 300 head, belonging to McIntosh of Westmoreland station in Queensland, arrived on the beef-butchering scene. McIntosh had come over 1,000 miles on this speculative venture and had lost half of his original herd due to an outbreak of redwater while in the Territory.[13] In March Nat agreed to hand over most of his share of the limited market to McIntosh, the result being that the price of beef rose by two pence per pound.[14] Possibly he felt that McIntosh deserved a break and that the dwindling business was not profitable enough to pursue.

The alluvial gold-bearing wash was shallow and soon worked out, and the reefs were mainly mere surface shows. When Coolgardie was discovered there was an exodus to the new fields and only about 100 people remained at the diggings, so the market for beef rapidly declined.

Queensland also experienced a slump in the cattle market and much stock was disposed of at boiling-down works. The Darwin, Pine Creek, and Burrundie goldfields markets in the Territory were limited and although a few mobs were sold there to the butchers Armstrong and Lawrie, it was insufficient.

With poor markets and a growing overdraft, Nat resorted to overlanding another mob of horses to the Territory. Making his usual fast passage from the eastern colonies, he had returned with fifty head to Burrundie, Northern Territory in September 1888.

After selling the horses and taking a quick trip out to Wave Hill he returned to Darwin in October with Tom Cahill and one of the Gordon brothers. The three boarded the SS *Guthrie* bound for Sydney.[15] The rain brought most operations to a temporary halt in the wet season until the country dried out enough to travel, and so this was the ideal time for him to visit his wife and attend to business affairs.

References

1. Map of Pastoral Claims in the Northern Territory of South Australia, 1883.
2. *Northern Territory Times & Gazette* 28/9/1878.
3. G.C. Bolton. 1953. *A Survey of the Kimberley Pastoral Industry from 1885 to the present.* C.E. Flinders. (nd). *Forty Five Years in the Great Nor-West of Western Australia.* unpublished ms.
4. There are conflicting accounts of the number of cattle initially taken from Wave Hill to stock Sturt Creek. In *Packhorse and Waterhole*, Gordon Buchanan says 1,800 but in another account credited to him he said 300 and the *Northern Territory Times and Gazette* stated 300. Perhaps 300 were taken initially followed by a further big mob as stated in the *Northern Territory Times and Gazette* 10/3/88.
5. Walkabout, 1/5/51 and A.S. Bingle. (nd). *This is Our Country.* Unpublished ms.
6. G.C.Bolton op. cit. Appendix 11. Letter from J.S. Foster from Denison Downs, dated 29/7/1890.
7. W.F. Buchanan. 1890. *Australia to the Rescue.* London.
8. Darwin Shipping and Passenger list, NT Archives.
9. Halls Creek Police Occurrence Books, entry dated 21/7/1887. AN5 Halls Creek, Acc. No.1422, W.A. Archives
10. G.H. Lamond, in Tales of the Overland suggests that Bluey was selling meat on the goldfields with P.J. Fox. Fox apparently bought R.S. McPhee's established butchering business in late 1886.
11. Halls Creek Police Occurrence Book, entry dated 1/10/1887. W.A. Archives, Acc. No.1422. *Northern Territory Times and Gazette*, 10/3/1888.
12. Halls Creek Police Occurrence Book, entry dated 3/11/1887. op. cit.
13. *Government Resident's Report of the Northern Territory and South Australia for the Year Ending 1888.* NARU DU392 .N76 1888.
14. Halls Creek Police Occurrence Book, entry dated 5/3/1888. op. cit.
15. Darwin Shipping and Passenger lists, NT Archives.

27. The Wagon Shed, Inverway Station
Buchanan Collection

28. First Denison Downs Homestead - 1890
W. Stretch, part owner
Battye Library, No. 78118
Published with permission of Mrs. H. Stretch, W.A.

15.

Wave Hill Crisis and the Search for Markets

1889 - 1894

As general manager of Wave Hill Nat inspected the property for a few weeks of every year. With Sam Croker now at Halls Creek he appointed Gordie as the new manager and Tom Cahill remained as head stockman. In January 1889, following leave in the south, general manager, manager and headstockman returned to the Territory together on board the S.S. *Changsha* and rode out to Wave Hill.[1] This was a time of crisis for the station. In the developmental years of Wave Hill the two major problems were attacks on stock by Aborigines and the lack of viable markets. While Gordie attended to the former, Nat addressed the latter.

When Gordie took over management he found the stock losses caused by spearing and disturbance of the herd were extensive. The Aborigines had two good reasons to wage war on the stock. As well as having developed a taste for beef they had a burning desire to get the whites off their land. Cattle were also easier prey than kangaroos so the spearing of stock became serious. Adult cattle usually didn't die when speared. Wounded, they were harried and chased until exhausted and further wounds could be inflicted. The stockmen were angered when they saw the slow agonised death endured by these animals and were further dismayed by the fact that cows in calf were prime targets because they were slow. The loss of breeding stock considerably delayed the herds natural increase. In order to minimise these attacks, frequent reconnaissances, boundary riding and punitive excursions were undertaken by the cattlemen. Despite these, the depredations continued and it was amazing that any cattle were held on the run at all.

A great limestone massif forty miles long lay to the west of the homestead and harboured a tribe of cattle-killers. Part of the valley between this range and the old Wave Hill homestead, about eight miles away, could be seen from the summits, and movements of cattle or stockmen were easily observed. From its precipitous ravines and gorges, honey-combed with caves and other well-watered hideouts, parties of war-painted cattle spearers made sudden raids to ravage the Wave Hill herd.

The spear heads in those days were of stone, and broke off easily. Many were recovered from just beneath the hides of 'killers' whose wounds had healed. Along the Overland Telegraph line Aborigines discovered that insulators made a good substitute for stone and thefts of these caused frequent interruptions to telecommunications. In later years glass, wire and iron were preferred materials. Tom Cahill was once hit by a spear, but as it was without a head the result was only a severe bruise.

The settlers found it necessary to employ Aborigines from other areas to assist on the stations so that loyalties were not divided. Between branding musters Gordie, two white stockmen and two Aborigines from the Katherine area, spent all their spare time patrolling the borders to the east, south and west. Pigeon Hole, an outstation of Victoria Downs, about fifty miles away to the north, acted as a buffer. On one scouting patrol to the south, the men made three dispersals from camps, twenty or thirty miles apart. In each camp there was evidence of freshly speared cattle. The patrol also found beef being beautifully baked in stone ovens. A hole in the ground was lined with limestone and heated, and in this was placed a quarter of a beast complete with hide. This was covered and sealed with paperbark and then a 2 foot mound of sand piled on top. The beef was a gourmet's delight - tender and juicy, and a welcome change from the stockman's regular diet of dry salt beef. The men took great satisfaction in eating some of their own beef prepared in this succulent manner. George was a wonderful tracker and if it hadn't been for him these camps would never have been found. Reputed to be able to track a ghost through a thundercloud, bare rock or basalt shingle merely slowed his pace, but did not defeat him. 'You no see'em track? Blackfella bin walk here.' Only a magnifying glass would have revealed any sign of a track to the whites.

It was usual for the Aborigines to scatter at the approach of white men, but on one occasion they didn't. Gordie said he had a lively recollection of several spears passing uncomfortably close. With Archie Ferguson, Mick Barry and George, they were tracking a fully armed party of thirty Aborigines. The patrol discovered them creeping up on some cattle at Red Lily Spring. When they saw the whites they separated into several smaller groups, moved off some distance and then stood their ground. This unexpected action forced the members of the patrol to separate too, the majority going after the larger group while Gordie was left to defend himself as best he could against seven or eight Aborigines. Two spear throwers were being fed with spears by their mates so Gordie fired a few warning shots hoping to scare them, then two rounds misfired. With the failure of his only means of defence he had to decide whether to confront or desert. The spears were coming alarmingly close, one passed through the horse's mane and another over its rump. Gordie kept his horse on the move to dodge the spears, while with a silent prayer he slipped another cartridge into the breech. The shot was good and succeeded in wounding one of the Aborigines in the side, which caused them to retreat supporting the injured man. Shots from a ravine alerted Gordie to the location of his mates so he rode in that direction and when he arrived found that they had scattered the rest of the band without causing any injuries.

The dry season days were bright and cloudless, with a nippy south-east wind. George found fresh tracks to the east and led the patrol onto another tribe of very surprised Aborigines. After crossing the downs and open plains of the upper Camfield, they moved ahead into the level forest country beyond. George followed diligently in the tracks and according to the nature of the tracking surface, averaged a pace of about four miles per hour. Often the patrol waited while he circled difficult terrain in order to pick up the trail on softer ground beyond. On finding the tracks he would pull up, look round at the men and typically, without a word, thrust out his chin to indicate the direction of the pursuit and they would again move forward. After two days of this patient manhunt the patrol found itself in a level open forest studded with striking rectangular outcrops of limestone. The blocks, which varied in length and breadth, rose precipitately from the forest floor and were about fifty to one hundred feet high. In Gordie's creative imagination they resembled prehistoric blocks of flats.

It was to this city of giant recumbent monoliths that the patrol came upon the illicit butchers. Feeling secure in their hideout, they were enjoying the spoils of the hunt. George pointed out a score or more hawks hovering and circling above the trees about half a mile distant. 'You see'em hawk? Blackfella bin camp there alright, got plenty beef too.'

The men rode slowly and carefully in the direction of the hawks. They kept the boys with the packs and spare horses well in the rear so that the rumble of packbags and the jingle of hobbles would not give them away. Until the patrol was within about 200 yards of the camp, the Aborigines were unaware of their presence. Some were awake, but the majority were sleeping off their meal. A horse snorted and a piccaninny cried out the warning, 'Cuddybah!' Spurring their horses into a gallop the men fired repeatedly with rifle and revolver at the dozen

or more figures rapidly escaping into the black cliff. Round the end of a narrow
tongue of rock raced the patrol, hoping to head them off in the grassy glade which
lay between it and a bigger hill. Shots were fired at some of the escapees seen
leaping for the shelter of the hill, and one was wounded.

Further pursuit was pointless, so they returned to the Aboriginal camp and
burned some stacks of spears and a few woomeras. This destruction of weapons
had the effect of slowing attacks on the stock because it took time for the
tribesmen to make new weapons. It was not the purpose of the Wave Hill patrols
to kill Aborigines but to inflict a wound as punishment. The Aboriginal women
and piccaninnies had melted into the bush in response to the attack and they were
not pursued because they were never the targets. It wasn't long before the
Aboriginal men recognised that their women were immune from punishment and
conscripted them as lookouts, and eventually they too took part in the cattle
killing.

Boundary riding the unfenced property to curtail and punish the depredations
and collecting stock that had been chased and scattered was an ongoing, thankless
chore which put a continuous strain on the resources of manpower and finance.
Gordie was a gentle, kindly man and completely unsuited for this type of duty, so
he was relieved when he could hand the job over. For the next two years Hugh
Gordon bore the responsibility and he too was glad to relinquish the job when
James Cullen succeeded him as manager. Cullen was a less scrupulous individual
in relation to the treatment of Aborigines, but as Gordie said, 'The history of Wave
Hill proved, violence begets violence.'

At the end of his year as manager Gordie rode to Darwin with Wattie where
they met Willie, Hugh and Mick Barry, who arrived by ship from Wyndham.
Willie Gordon was returning south after having delivered a mob of cattle from
Queensland to Lillamaloora in West Kimberley. The five men boarded the S.S.
Airlie on 21st November, 1889, bound for Sydney.[2]

C.B. Fisher first experimented with shipments of cattle to the East in 1886,
with the intention of finding a market for surplus fats from Fisher & Lyons
properties. H.W.H. Stevens, their general manager, visited Singapore and Java in
1887 and reported that there were good prospects for a live beef trade in those
countries. Land was acquired and fencing and shipping facilities were set up near
Darwin in preparation for this trade.[3] Early in 1889, when the Singapore market
looked promising, Hugh and Wattie delivered 415 Wave Hill fats to Darwin,
which brought fifty shillings per head. However this market was short lived, so
Nat continued to investigate other market opportunities.

By 1889 the local market in the Northern Territory was saturated for the first
time with Territory beef. The initial boom in the industry was followed by a
slump which sent many lessees to the wall. There were a multitude of reasons for
this. The astronomical costs of setting up, high wages, cartage and materials costs,
Aboriginal depredations, redwater fever, lack of and distance from markets,
unreliable water supplies on stations and the atrocious condition of stock routes

combined with uncertain surface water for travelling stock.

Fisher & Lyons properties had changed ownership and were in the hands of English shareholders, namely the North Australia Territory Investment Company. The English shareholders did not understand the country and the conditions and in a very short time lost heart and sold their Territory properties to Goldsborough, Mort & Company. Stevens continued to work for these organisations and to push ahead with the cattle-export service to give pastoralists a market.

While Nat was striving against mounting odds to make Wave Hill prosper, W.F. Buchanan tried to emulate Fisher & Lyons and dispose of the Territory leases and stock to British shareholders. He visited England for this purpose in 1887, but was unsuccessful.[4] It is highly improbable that this action would have met with Nat's approval and it is possible that he was not even consulted.

It was now a matter of urgency that a market be found. Nat with his usual determined optimism investigated three market opportunities in 1890. The records indicate that he took 350 head via the Fitzroy River and Ninety Mile Beach to the Nullagine goldfields, personally investigated the Singapore livestock market by accompanying thirty head aboard ship from Derby, and then sent a trial shipment of sixty head from Derby to Perth.[5] These were all Wave Hill cattle from the Sturt Creek leases brought over by Gordie.

No information appears to be available regarding the result of the Nullagine venture so it is assumed that it was a failure. The Singapore trip was a financial disappointment because the cattle shipped at a cost of five pounds per head sold for only eight pounds. Nat, did however, gain first hand information about that market and reported to W.F. Buchanan that to make it successful they would require a resident agent and land for paddocking stock. W.F. believed this proposal was too costly to risk single-handed, so he wrote to the South Australian Government asking for support for the project in the form of a subsidy. Eventually the Government approved a subsidy of two pounds per head and the steamer *Darwin*, with a capacity to carry 250 head, was built especially for the live beef trade.

The third venture was more successful. Despite sixty of the original ninety head shipped dying as a result of the primitive shipping facilities, the remainder sold at Fremantle for sixteen pounds per head. A great improvement on the price offered in Singapore! Fortunately Nat had the foresight to insure the cattle at ten pounds per head so his losses were covered. This promising market eventually proved of most benefit to those pastoralists in West Kimberley because they had ready access to the port of Derby. East Kimberley and Territory cattlemen were at a decided disadvantage because they had to drive their cattle long distances to the port.

Four experimental cargoes were shipped from Darwin to Singapore in 1891 with favourable results. Wave Hill supported the project by sending 600 cattle to the new market. The trade with Singapore was officially established in April

1892, and Goldsborough, Mort & Co. set up an agency in Singapore and maintained a wholesale and retail butchery business there. The following year a fall in the rate of exchange made the Singapore trade unprofitable again and the shipping service negotiated a three year contract with Batavia.

Mistrust of Goldsborough, Mort & Co. plus the losses of travelling stock due to redwater made many pastoralists reluctant to support the export trade. The contractor, represented by Stevens, was obliged to employ his own drovers to bring the stock in from the stations. Only the assistance of the Government subsidy enabled him to pay fifty shillings per head on the property and take all the risks and losses.

Between the official inauguration of the export service and the end of 1894, Wave Hill was one of the cattle export service's main supporters, shipping a total of 1,173 cattle to Eastern ports. Unfortunately this was not a big enough market to absorb the growing number of fats on the property. Problems such as weight loss of stock during shipping, mismanagement by Stevens, redwater fever and the fall in the value of the Strait's dollar plagued the shipping service. The cattle tick eventually sounded its death knell in 1897, when Java prohibited the importation of stock from the Northern Territory.

The year 1891 saw Kate finally reunited with her family in the Kimberley. Four years earlier Gordie had taken up pastoral leases east of Halls Creek which he named Flora Valley.[6] This was to be Kate's temporary home until Nat had found a reliable market for his cattle and they could settle permanently on Wave Hill. Obviously Nat was of the opinion that sooner or later his Victoria River property would be viable.

In March 1892, when all other markets had only partially met their requirements, Nat, Hugh and Wattie made a determined attempt to establish a stockroute to southern Western Australian outlets. The dauntless trio decided to overland 1,000 cattle off the Elvire block, and drive them some 1,500 miles to the Murchison goldfields.

This was Nat's last big droving trip. It was another epic, trail-blazing journey where his endurance, superior skills as pathfinder, bushman and drover were to be tested to their limits. For a man in his mid-sixties this was such a tremendous undertaking that people today shake their heads in awe and disbelief. If Nat felt his many years of hardship, he never mentioned them. Kate, knowing better than to try and dissuade him from his course, accepted the inevitable, and as in earlier years she and Gordie faced the months of silent waiting and worrying together.

The droving party was made up of Nat, Hugh and Wattie, stockmen Bob Carney, Edgar O'Donnell and Jim Boyd with the wagon and team, plus a couple of Aboriginal stockmen. They took the route west from the Elvire to the lower Fitzroy River, reaching the west coast at a point between La Grange Bay and the Fitzroy. The cattle were separated into two mobs travelling a day or two apart.

Ahead of them along the coastal stretch lay five hundred miles without surface water, until the De Grey River. The track had been in use previously but only rarely, and the government had dug wells fifteen to twenty miles apart and equipped them with buckets and troughs. Typically, some of the wells had filled with sand, and the troughs and buckets had been either stolen or fallen into disrepair. The water supply which would have served a small plant of horses was quite inadequate for such a large mob of cattle.

The three old stagers came prepared for all emergencies, carrying canvas troughing sufficient to water forty head at a time. The cattle were unused to drinking from a trough and initially were too timid to come close, but after a dry stage or two their thirst gave them courage and they drank their fill. Drawing water by hand was time consuming and labour intensive, and the men, unable to keep up the supply to the thirsty cattle, were forced to reduce the length of troughing. Now they could supply only ten beasts at a time but at least they could physically keep the water up to them. It took about seven hours to water the mobs and very often there was not sufficient for all, so some went for a day or two without a drink. The horses and plant were catered for by sending them on ahead. When a good supply was struck, they resorted to the methods of seasoned campaigners and camped for a day or two, while they investigated the mysteries of the country immediately ahead.

Feed was plentiful and as there were no other cattle in the area, the herd was unsupervised at night. When camped one night before reaching the De Grey River, something disturbed the cattle and a number broke away. The next day Hugh Gordon and one of the Aborigines set out to recover the escapees. Hugh tracked them across forty waterless miles of sandy country before being forced by thirst to return empty handed. Fortunately it was winter time, otherwise they would never have made it because the horses were badly in need of water and quite done in. The lost cattle fared well and they were eventually recovered watering on the De Grey River.

From the De Grey, surface water was available so the two mobs of cattle were united as they turned inland to negotiate a stretch of 120 miles via Marble Bar to Nullagine. Their real hardships started after the Nullagine River. It was the driest time of the year, there were no wells and the surface waters were unknown to the drovers. Packhorses were substituted for the horse-drawn wagon to enable greater mobility. Between Roy Hill and the Ashburton River they knew they would be in dry country so at Emily Springs, the last known permanent water, they prepared for a fifty mile dry stretch. Nat, a master of bushcraft, looked for soaks in the bends of the dry rivers, and if he could locate a patch of damp sand under the surface he would test the depth with a stick. Experience told him whether there would be enough water there for the cattle. If so, then the soak would be scooped out, logged and bushed and a man could stand in it to hand out buckets of water to another on the surface, who would pour it into the canvas trough.

In his later years Nat took to carrying an umbrella to protect him from the sun.

This practical eccentricity became part of his legendary image. The horses he rode had to be accustomed to this rather frightening apparatus and a lot of thrills and spills were had by the stockmen and boys doing the breaking in. If the umbrella was unfurled too quickly it would startle the horse, and the exhibition of buck jumping that followed provided plenty of amusement for the bystanders. Now Nat found another practical use for the famous umbrella. By this time the bullocks were very quiet so there was always the risk that in their desperation to get a drink they would crowd the trough and trample it. To prevent this happening Nat would stand by with the trusty gingham and if the cattle became too eager he would rush at them, while vigorously opening and closing the umbrella. This never failed to have the desired effect.

The cattle travelled well because there was plenty of feed along the route, but finding water was a constant problem. A good passage was made through the Opthalmia and Hamersley Ranges which lay between the Fortescue and Ashburton Rivers.

A two day dry stage followed so Nat and one of the Aborigines rode ahead about sixteen miles and located some surface water, just enough to give the cattle one drink, but no more. The Aborigine was sent back to tell the drovers to move the cattle up and the half-perished mob started on the road in the moonlight and reached the water at 10 a.m. next day. This water provided only temporary relief - more water was desperately needed. Nat and Wattie rode ahead with the cattle following in their tracks. They were at the point of no return and there was no escaping the stark reality that they must find water very soon or the cattle would perish. Travelling down the Ashburton the cattle sensed the proximity of water and ran on looking for it. It became essential to control the pace of the line of thirsty beasts or risk losing them before water was discovered.

At the height of this tense situation, word came back from Nat that he had located a waterhole to suit their needs. This dry stage had involved both day and night travelling and even the men ran out of water. The men and the mob reached the hole thirsty and exhausted and the cattle were sent on to drink in small lots. The relieved party recuperated by the waterhole for a day, reassured by the knowledge that from here on the going would be easier. From a tributary of the Ashburton they crossed to the Gasgoyne and finally came to Beringarra on the upper Murchison River.

> *Messrs Buchanan and the Gordons got through with their cattle to the south, Messrs. Darlot Bros. buying them:...*
>
> *Northern Territory Times & Gazette, 10/3/1893.*

The cattle arrived in surprisingly good condition thanks to the good feed along the route. They were counted over to the Darlot Brothers and held in a sheep paddock. That night they rushed and flattened the fences so the Gordon brothers were forced to delay their return to the Elvire with the horses and plant by three weeks, while a suitable bullock paddock was constructed. The cattle brought eight pounds per head when delivered to Perth by the Darlot brothers. Hugh and Wattie arrived back at the Elvire with the horses and plant in April 1893, having taken just over one year for the round trip. The new droving track blazed from the De Grey to the Murchison became known as Buchanan's track.

To encourage the pastoralists of Western Australia to produce enough meat to supply the local market, the Tariff Act of 1888 placed a protective duty on all livestock imported to Western Australia for slaughter. The tax was thirty shillings per head for cattle and one pound per head for horses. As Nat planned to supply the next mob from Wave Hill and would be eligible for the tax he negotiated with the Western Australian Government in Perth to pay after delivery. In an optimistic mood following this satisfactory arrangement Nat sailed north to prepare the next draft.

Interest in overlanding cattle to this promising new market came from as far away as Queensland. In anticipation of an influx of stock to supply the southern goldfields market the Government renovated all the wells along the track. However, the Western Australian cattlemen wanted to avoid competition and retain their monopoly of the markets, so they petitioned their government to block cattle from other States. The Stock Act of 1893 extended the tax to include all imported cattle with the exception of stud stock.[7] Loopholes in the Act were thus closed to prevent cattle from outside Western Australia's borders being bought in as store stock and fattened before being marketed for slaughter.

Unaware of this new development, Nat confidently returned to Wave Hill and immediately started mustering 2,000 - 3,000 steers in readiness for the Gordon's return with the plant. The cattle were no sooner on the road than he was notified from Halls Creek that the Western Australian Government had altered the Act, effectively closing its border to Queensland and Northern Territory cattle.[8] This news was a crushing blow to Nat. Relying on his negotiations with the Western Australian Government, he had made quite a financial commitment to get the cattle mustered and on the road.

There was nothing to do but ride with the punch and salvage what costs he could, so he mustered 900 steers from the Elvire and set them on the road for the Murchison, under drover W. Richardson. This trip to Cue on the Murchison took six months and W. Richardson was accompanied by his wife, a Malay, and three or four Aboriginal stockmen. Early in the trip Richardson was injured by a fall from his horse and was unable to do his job, so Mrs Richardson took over while he rode in the wagon. Dressed like a man, she undertook all the duties of a boss drover and delivered the cattle safely to W.F.'s agent, Charlie Buchanan. This must certainly have been a classic example of a pioneer woman's courage, loyalty and endurance.[9]

In September 1892 while Nat was droving to the Murchison, his friend and travelling companion of many years, Sam Croker, was shot dead.[10] The forty year old Croker was acting manager of Auvergne Station in the Territory at the time and his murderer was a man named Charlie Flannigan. Flannigan was of mixed descent and it appears he may have come across to the Territory from Richmond Downs in Queensland with the 20,000 head of cattle in 1881-2. Gordie said that Charlie once told him that Old Bluey was the only man he could work for:

> *Alternately morose or jocular, he seemed to feel generally the inhibitions of his caste. He was a fair horseman and stockman, and an expert in the drafting yards and branding pen. Illiterate, but fairly well spoken, he seldom swore. One day on a droving trip, while riding slowly backward and forward behind a big mob of cattle, he said to me, 'Looking at these cattle gives me a pain in the neck' - a prophetic comment.[11]*

One night, presumably in response to one of Sam's remarks over a game of cards, Flannigan rose silently from the table and went out only to return a few minutes later with a rifle. In front of witnesses he shot Croker first in the chest and then finished him off with a shot to the head. After some indecision he took Croker's horses and gear and avoiding all roads and intermediate stations, struck through the bush for about 150 miles to Ord River station. There he told the manager, F.C. Booty, the whole story. Somehow, Booty and Jack Kelly the head stockman persuaded Charlie to take the horses and give himself up to the police at Halls Creek. Jack Kelly offered to accompany him and introduce him to the police. It was a dangerous undertaking requiring infinite tact and cool courage, for Charlie, always irresponsible and impulsive, was armed and might regret his decision at any time during the 120 mile ride. At Halls Creek Charlie gave himself up to the police and was eventually tried in Darwin, and sentenced to death. He was executed at Fannie Bay Gaol and holds the dubious honour of being the first man hanged in the Northern Territory.[12] While awaiting execution Charlie did many drawings depicting the outback life he had known. They were of the stations where he worked, cattle camps, horses etc. all done from memory. Now they provide a valuable pictorial record of some of the early stations. These pictures are also important because they are an example of work done by an Aboriginal artist in the European style and it seems, uninfluenced by traditional Aboriginal art. Perhaps Charlie Flannigan never experienced a traditional lifestyle and only identified with his European heritage. Sadly he was probably never accepted as an equal by the white community.[13]

Sam Croker was buried on Auvergne and a few years ago his family restored his grave.

With an excess of fat cattle, a lack of markets and supposedly the prospect of the banks foreclosing, W.F. Buchanan put Wave Hill and the Elvire block up for auction in Melbourne in February 1894.[14] Nat did not agree to the sale and refused point blank to sign the sale papers presented to him by his brother's agent in Sydney. It is reported that he clapped his hat on his head and walked out of the office in disgust. Despite not having his partner's signed agreement, the wealthy older brother proceeded with the sale. At auction both properties fell to W.F. Buchanan - Wave Hill for the purchase price of 15,000 pounds.

Donald McIntyre of Dalgonally, Nat's confidante in this matter, said that his old friend was very angry over what he considered an unwarranted sale.[15] Despite receiving advice from McIntyre and others that he could legally remain a partner, and knowing that he could substantiate the validity of the partnership, Nat refused to take legal action against his brother. Even if he could afford the legal fees involved, which is doubtful, airing the family's dirty linen in public was not Nat's style. These two facts probably ensured that W.F.'s strategy would succeed without tarnishing his reputation.

The true reasons why W.F. Buchanan put Wave Hill on the market will probably never be known. The fact that he purchased it at auction himself seems to indicate that he wanted Nat and the Gordons out so that he could assume full control. Nat was a hardy, old battler who was determined to hang on to Wave Hill at all costs and so would have resented W.F.'s attempts to sell the property in 1887. Nat knew the country and the conditions while W.F. didn't. For these reasons it seems highly likely the partners would have had differences of opinion over management. If William's reason for selling was genuinely because he couldn't service his share of the debt he was still only entitled to sell his own share in the property and this would not have nearly covered his losses. Without any inside knowledge one can only draw the conclusion that W.F. was confident that one day Wave Hill would prosper and he would get his money back.

Both Nat and W.F. would have been aware in 1892 of the proposed new shipping service at Cambridge Gulf.[16] It must have been a light at the end of the tunnel for both of them. Nat would have been determined to wait for this opportunity to eventuate. In the mean time the slump in the market meant that W.F. could gain full control of the property for a pittance. In the absence of information to the contrary and however charitably one wishes to view W.F.'s. action, it seems apparent that he was determined to be the sole owner of Wave Hill and was prepared to go to extraordinary lengths to do so.

Nat and the Gordon brothers were cheated out of their rightful share in Wave Hill's eventual prosperity. Their financial contribution and hard work for a period of more than ten years went unrewarded. It seems that Nat confided his true feelings to Donald McIntyre and thereafter remained silent about the matter, thus ensuring that his brother's reputation was never in question. Once the sale was over, Nat closed this chapter of his life and rarely spoke of it again.

Because there was no money, wage claims were not met. Nat was entitled to 100 pounds per year for his salary as General Manager, but received 150 young steers in lieu. Gordie and the Gordon brothers accepted a total of 600 cattle which had been over-valued at two pounds per head. At that time young heifers were available from Victoria Downs for one pound per head. All these cattle, including

Nat's, were moved to Flora Valley and the Gordons then formed a partnership with Gordie.

Following Nat's death, Kate and Gordie were encouraged to take W.F. Buchanan to court over the irregularity in the sale of Wave Hill. Gordie said that a prominent King's Counsel was so confident he could win the case that if he lost he was prepared to forego the legal fee. Charlie Buchanan is reported as commenting that his uncle was confident that he would not have to face the courts and if he did, was prepared to pay 70-80,000 pounds settlement. Respecting Nat's wishes, Kate refused to take court action against her brother-in-law and eventually accepted an out of court settlement which was much less than full equity. The fact that W.F. was prepared to make any settlement at all casts a cloud of doubt over the legality of the sale.

In the twenty nine years he was associated with Wave Hill, W.F. Buchanan never visited the property. He appointed the very able Tommy Cahill as manager to succeed Jimmy Cullen who moved to Flora Valley. W.F.'s nephew, Charlie Buchanan, took over Nat's job.

After the jetty was completed at Wyndham in 1896, Connor, Doherty and Durack inaugurated a cattle shipping service which served the East Kimberley and neighbouring Territory runs. This access to southern markets combined with the rise in cattle prices saw Wave Hill go on to great prosperity. Tommy Cahill remained as manager until about 1905, when he went into partnership briefly with Gordie and the Gordon brothers on Gordon Downs. In 1912, after W.F.'s death, the station was sold to Charles Hadfield Wright for 200,000 pounds. It changed hands again five years later when Vesteys became the owners and they held the property until July 1992, when it was purchased by Queensland grazier Mr Brian Oxenford. The sale price was not disclosed but indications are that it was in the vicinity of eight million dollars.[17]

Wave Hill, the oldest cattle station in the Victoria River District, still thrives today. An area of 32 square kilometres was surrendered for a Public Pound in 1923 and is the site of the present Police Station. The old homestead was washed away in a flood in 1924, so a safer site was selected for the new station, built a year later. These buildings had deteriorated so much by the mid-sixties that another homestead was constructed on a different site.[18]

Other changes to the property came about when land was resumed for the use and later freehold title of Aboriginal people. In 1948, Hooker Creek - originally Catfish - and now Lajamanu, was surrendered by the station. In 1966 the Gurindji people, led by elder, Vincent Lingiari, staged a walk-off and went on strike for better pay and conditions for Aboriginal people working on Wave Hill. They camped on Wattie Creek near the site of the original Wave Hill homestead. This strike extended into a claim for the return of traditional Aboriginal lands and represented the beginning of the Aboriginal land rights movement. The land, west of the Victoria River and north of Hooker Creek, was officially granted in 1975 and the camp on Wattie Creek became the settlement of Daguragu. Wave Hill applied to subdivide and in 1984 became Wave Hill and Cattle Creek Stations.

References

1. Darwin Shipping and Passenger lists, NT Archives
2. Darwin Shipping and Passenger lists, NT Archives.
3. All details of the Darwin cattle shipping service were obtained from the *Government Residents Reports of the Northern Territory of South Australia* dated, 1890, 1891, 1892, 1893, 1894. 1895, 1897.
4. Ross Duncan, 1967, *The Northern Territory Pastoral Industry 1963-1910*, Melbourne University Press.
5. G.C. Bolton, op. cit. and C. E. Flinders. (nd). *Forty Five Years in the Great Nor-West of Western Australia*, unpub. ms.
6. C. Clement. 1993. *Kimberley District Pastoral Leasing Directory 1881 - 1900*, National Heritage.
7. G.C. Bolton. 1953. *A Survey of the Kimberley Pastoral Industry from 1885 to the Present.*
8. Several years later the border was closed to protect Western Australia from redwater fever which was prevalent in the Territory.
9. C.E. Flinders, op. cit.
10. *Northern Territory Times & Gazette*, 24/2/1893.
11. When Charlie Flannigan was awaiting execution in Fannie Bay Goal he did some sketches of stations where he had worked and wrote the name of each station on the drawing. This indicates that he had some rudimentary education and was not entirely illiterate.
12. *Northern Territory Times & Gazette*, 14/7/1893.
13. Andrew Sayers. (nd). *Aboriginal Artists of the Nineteenth Century*. Oxford University Press, Melbourne.
14. *Northern Territory Times and Gazette* 22/2/1894
15. Donald MacIntyre, Statutory Declaration, 13/1/06. Buchanan Collection.
16. *Northern Territory Times and Gazette* 15/7/1892
17. *The Weekend Australian*, July 18-19 1992.
18. W.A. Low, B.W. Strong & L. Roeger. September 1986. *Resource Appraisal of Wave Hill Pastoral Lease 911*, prepared for the Conservation Commission of the Northern Territory, Alice Springs.

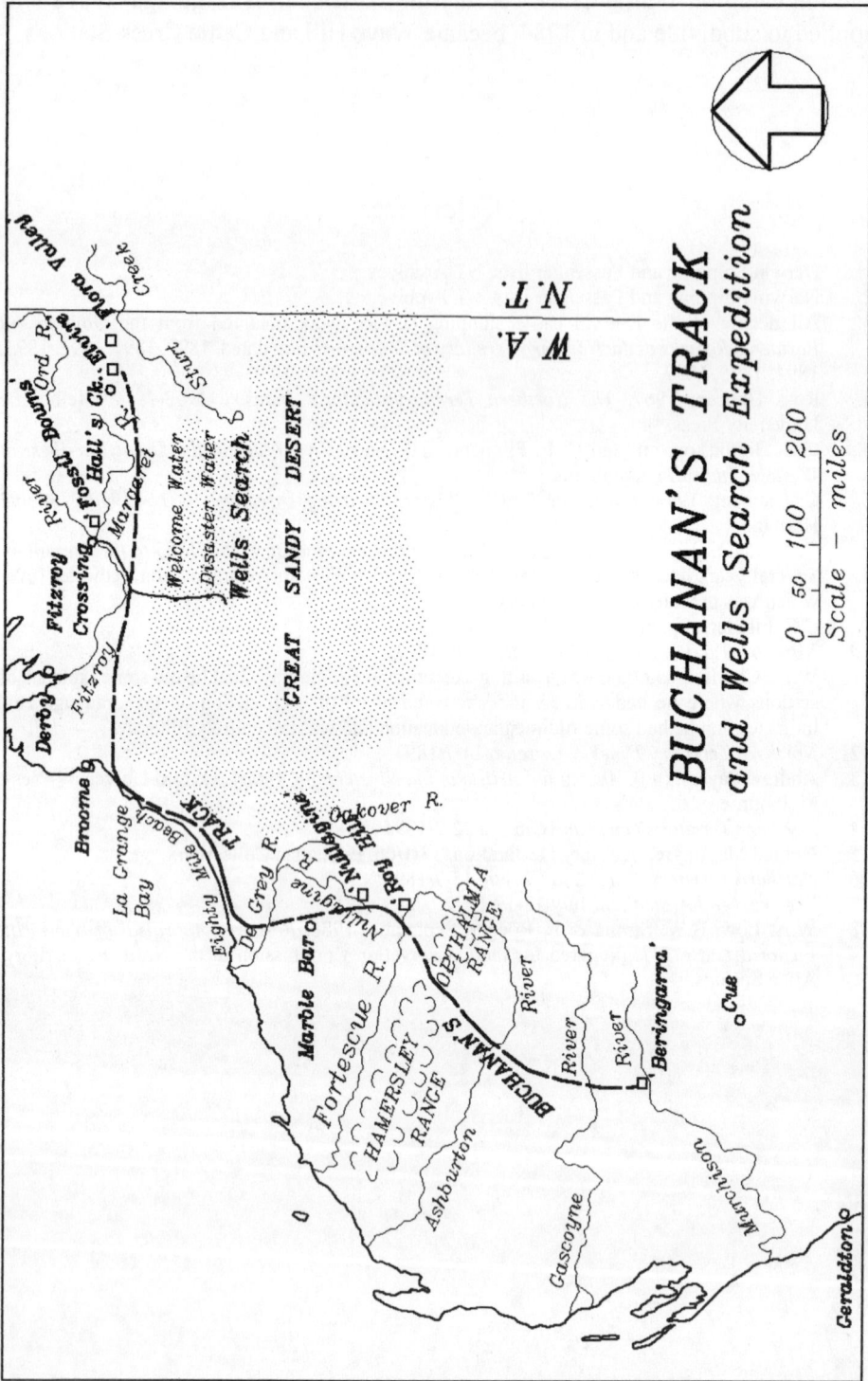

Buchanan's Track and the Wells Search

29. Hugh (left) and Wattie Gordon on the steps of Flora Valley Homestead
Battye Library, No 7808B. Published with the permission of Mrs. H. Stretch, W.A.

30. The Grave on Auvergne Station of Sam Croker
Photograph courtesy of Darrell Lewis

16.

Kimberley Cameos

1891 – 1896

In September 1891, Nat and his wife arrived in Wyndham en route to Flora Valley. They had travelled from Sydney to Port Darwin on the S.S. *Guthrie* and there boarded the S.S. *Rob Roy* for the Cambridge Gulf.[1] From this torrid new settlement they set out by buggy and pair for their son's property. It had been thirty years since they had made their honeymoon trip from Port Denison to Bowen Downs. It must have brought back many memories of earlier days when in a similar mood of optimism, they faced the future together. This was to be the first step towards taking up permanent residence at Wave Hill and would have been their first shared home since the early years of their marriage.

After the comparative civilisation of New England and Sydney, Wyndham must have seemed an alien place to Kate. In 1886, at the height of the gold rush, when Governor Broome declared the town of Wyndham it became the main port for supplies to the goldfields and for the cattle stations of the interior. It was established for several reasons. Firstly, the South Australian Government were thinking of making a port on the Victoria River, just across the Territory border, to capture the goldfields market and to provide an outlet for stock. Secondly, the prospectors were arriving at Cambridge Gulf in large numbers and avoiding customs duties. Thirdly, pastoralists and shipping companies had petitioned the Western Australian government for a port.[2]

In a period of four months at the zenith of the gold rush, nineteen vessels arrived in the port carrying 1,700 passengers and over 1,000 horses. Most of these men and horses made their way 200 miles inland to try their luck at the Halls Creek diggings. All manner of men were pouring into the north-west by sea, and overland from east and west. Ten thousand were reported to have passed through the Kimberley fields between 1885 and 1887, but only a few found gold.

After a rough, dusty and tiring journey travelling through the spectacular Kimberley landscape, Kate and Nat arrived at Flora Valley. The flat to undulating downs of Mitchell grass interspersed with belts of scrub that greeted them were reminiscent of Bowen Downs. From the time of her arrival Kate cast her gentle presence over the station. The rudimentary cottage with a palm leaf roof, situated on a grass plain back from the lovely fig-lined Elvire, responded to a woman's touch to become a home rather than a camp. Kate brought comfort and brightness to the austere, all-male environment. Despite few amenities of civilisation, all passing travellers were welcomed and thus news from the outside world reached secluded Flora Valley. No matter what the hour, Kate would put on the kettle and lay the table, and share the rough station fare with her guests. Attired in the full length dresses with high neck and long sleeves of the period she must have felt uncomfortably hot, especially in the wet season. The pioneer women not only had to put up with the confinement and discomfort of their dress but also with the risk of their skirts catching alight when cooking over a camp fire. With Nat always coming and going and her son and brothers often away mustering or droving she frequently had only the Chinese cook, a gardener, and the Aborigines who camped down at the creek for company. Kate was content and at one with the solitude of the bush and happy in the knowledge that she was at last able to make a worthwhile contribution to her family's welfare.

Flora Valley was much closer to civilisation than Wave Hill. The shanty town of Halls Creek was only sixty five miles away and had two hotels, a telegraph station and post office. Charles Danvers Price, who was the first Government Resident in Wyndham became the first warden of the Kimberley goldfields, arriving on the fields in September 1886. His old position in Wyndham was taken up by H.H. Hare. One of Price's initial tasks was to select the camp for the police on Elvire Creek. Later in March 1888, this camp was moved to the old Halls Creek townsite. From 1886 until early in 1887 the mail runs between Derby and Halls Creek and Wyndham and Halls Creek were handled by the police. Price had a hot temper but was capable and so well-liked by the majority of the miners and businessmen that when he left the goldfields in July of the following year they presented him with a gold watch. He returned to Wyndham as Government Resident. His successor, Warden Finnerty, arrived on the fields in June and combined the roles of magistrate and warden.[3]

For a short time, during the height of the rush, a doctor by the name of Langdon set up a hospital tent on the fields near McPhee's Gully, some 20 miles from the police camp at the Elvire.[4] Common illnesses among the fossickers were dysentery, scurvy and fever. These, plus the occasional murder and accident, were the main causes of death on the fields. When Kate arrived in the Kimberley there was no doctor to service the district so she and her brothers nursed many people who fell ill.

There was a monthly packhorse mail service to Flora Valley, and the arrival of the mail contractor with a delivery of letters and papers was always something to look forward to. In 1891-2, the contractor for the delivery of mail between Wyndham and Halls Creek was Donald Swan, one of Gordie's droving mates on the Ord River trip. After arriving with the first cattle to Ord River Station he worked on the property for a time, then started packing supplies between the goldfields and Wyndham.[5] Phillip Watts, nicknamed Jack, was another mailman. He held the first mail contract between Fitzroy Crossing and Ruby Creek, while C.E. Flinders had the first contract between Derby and Fitzroy Crossing.[6]

In a letter to Gordie, Jack related the story of the time Nat rescued him from perishing between Halls Creek and Wave Hill. '...your dear old father the day he picked me up at Sunbeam, when I nearly went west from thirst, when en route to Wave Hill. I can see him yet in the lead of the party heading straight as an arrow for Flora Valley waving a twig in front of him as he rode along'.[7]

Fragile articles got rough treatment in the mail contractor's packbags. They were dumped on the ground at least once each day for a fortnight and hammered in the middle to make a groove for the surcingle. This was not conducive to good delivery. One month the Flora Valley mail bag contained the powdered remnants of Tom Cahill's wedding cake distributed among the papers and letters. The badly damaged package was addressed to Tom's brother, Matt, at Gordon Downs. The Flora Valley crew took great delight in patching up the package and replacing the original contents with station 'brownie'. This mixture of sweetened damper with a few raisins added was tough enough to survive all the rigors of packhorse travel. The package was bound with a neat satin ribbon and dispatched to Matt in the next load of station supplies.

'Cripes, what's this?' said Matt to his mate Paddy when he received the package.

'Looks like your brother Tom's wedding cake.'

Matt looked at the writing on the package, 'It's Tom's writing alright, let's have a look at it.' He cut the ribbon with his pocket knife and opened the pack, 'Well I'll be damned.'

'Doesn't look up to much for a wedding cake' chuckled Paddy.

'Up to much!' exclaimed Matt. 'I could make a better cake than that myself'.

Flora Valley's nearest neighbour in the very early days was Fred Booty, on the lower Elvire. Booty came to the Kimberley as bookkeeper on Ord River when Button was manager. He was a nephew of W.H. Osmand and later became one of Button's many successors in the manager's job. His property was called Koojubrin, the name being the result of his frequent request of travellers: 'Next time you are passing could you bring...'. Wattie Gordon had a phenomenal memory for cattle and once when he found a weaned, unbranded bull among the stock in the Flora Valley yards, he realised it didn't belong to them and so sent an Aboriginal boy to Booty with a note describing the beast. In two days his reply was received written very neatly, but short and to the point. 'The red micky is mine!'

Initially Flora Valley's neighbours were few and far between. Argyle, the home of the Duracks, was over 100 miles to the north. Gordie first met the brothers, M.P. and J.W. Durack and their father Patsy, in November 1886, soon after the delivery of the first stock to Sturt Creek. They had a temporary camp on the east bank of the Ord which consisted of a tarpaulin tent and bough shed. This was later replaced by a grander homestead on the banks of the Behn River. Today the site of the old homestead lies beneath the mighty Lake Argyle, but the house has been preserved for posterity on another site.

> *4,000 head of cattle have arrived at Hodgson Crossing on the Roper en route for Kimberley district M & J Durack owners, in charge. They will camp at the Roper till the end of the wet season.*
>
> *The Northern Territory Times & Gazette, 28/2/1885.*

The Duracks acquired leases in the Kimberley and Northern Territory, left their home in western Queensland, and set out on the long trek north. M.J. Durack and his brothers, with Hayes and Tom Kilfoyle, took the inland track to Cresswell Creek where they turned due north, struck the head of the Limmen River, and followed it down until they reached the Gulf Track. Their father, Patsy Durack, went around by boat and was followed at a later date by his wife and daughter. The stations Bullita, Ivanhoe and the Twenty Mile (near Wyndham) were eventually formed and when the firm of Connor, Doherty and Durack was established Newry and Auvergne were incorporated. They were mainly stocked with cattle from Wave Hill delivered by Charlie Buchanan from W. F. Buchanan's Elvire block in 1894. In Gordie's words:

> *There are stockmen who are poor managers, and managers - good ones - who are poor stockmen. 'Long Michael' Durack of Lissadel was, like Tom Cahill, expert in the dual capacity. To meet him in the city, swinging his cane like a roistering swaggering blade, one would not suspect that behind that demeanor was a dynamic stockman and a kind and hospitable personality.*

Another member of the Durack clan, Mr. J.W. Durack, recalled riding with Nat at the Brockman. He was fascinated when for lunch, Nat dug deep in his pocket for a few raisins. He liked to travel light and he enjoyed dried fruits so this became a habit of his. For a quick meal or snack eaten on horseback, a handful or two of dried fruit in his pockets provided the perfect answer. Dried apples were his favourite because when followed by a drink of water, they 'Quite filled him up'. It was said that this veteran drover could last without water for longer than a camel! An exaggeration, no doubt, but Nat was known to prepare his constitution for dry stages by limiting his water intake and he explained his survival technique to Durack that day on the Brockman. After a dry stage when others gulped the precious liquid Nat, supposedly, would just rinse out his mouth.[8]

Patsy Durack's wife died from malaria in 1893 and in her book *Kings in Grass Castles* Mary Durack records the Buchanan's visiting Argyle to pay their respects.

> *Nat Buchanan and his wife who had come to Kimberley not long after Grandmother drove by buggy all the way from Flora Valley. Grandfather took his old friends' hands, thanking them wordlessly, and turned to talking of the horses in Buchanan's plant, of a fine colt one of the station brood mares had got from the stallion Sultan that had near been taken by horse thieves. But his effort at control had been too great and in leading his guests to the verandah he had reeled and fallen in a dead faint. Buchanan and Father lifted him to a bunk where he quickly recovered, made his apologies and went on with talk of everyday affairs.*

The MacDonalds were another early Kimberley pioneering pastoral family. They acquired leases on the upper Fitzroy and Margaret rivers. The brothers Charlie and Willie set out from the Goulburn district of New South Wales in 1883, with a small mob of cattle and a bullock wagon. Drought struck them in Queensland and their small herd was decimated. Determined to reach the Kimberley, they took jobs and replaced their lost stock, then pushed on through the Barkly Tableland and so down the Limmen River to the coastal track. At Battle Creek in the Victoria River district, both brothers went down with malaria, Charlie was so badly stricken that he was forced to return to Katherine with a group of prospectors. Although he had the help of a good bushman called Edmonds, Willie was left to complete the harrowing journey short-handed, and low on rations. Realising from the cattle tracks that there was a party close up ahead, Willie set out on horseback to overtake it in the hope of securing some provisions. He came upon the Duracks camped on the Victoria River and was able to borrow enough supplies to make it to the Ord. Charlie recovered and rejoined the cattle before the journey's end. The brothers won through to the Fitzroy with less than 500 head of cattle and established Fossil Downs. This droving trip from Goulburn to west Kimberley took two and a half years and was one of the longest on record. It took the MacDonalds some years before they built up their herd sufficiently but they battled through and the property still remains in the hands of their descendants.

Close Flora Valley neighbour, William Henry Stretch, and his partners Foster, Weeks and Lewers, overlanded cattle from Queensland in 1888 and settled on Sturt Creek about forty miles below the Button hole. They named the property

Denison Downs and stocked it with breeders bought from Frank Hann of Lawn Hill Station in Queensland. At that time it was the only occupied holding on the 300 mile length of Sturt Creek. Stretch was an imposing figure standing 6' 3" tall. Well read, his intellect matched his impressive stature and personality. Nicknamed the 'professor', because of his knowledge of books and his scholarly leanings, he was a great conversationalist with a remarkable memory, a sharp intellect and a straight-forward manner. A genuine humanitarian, Stretch was well respected and troubled men sought his advice, encouragement and understanding. The 'professor' also brought the classics to the culturally starved of the outback. Gordie recalled:

> One cold winter night in the original slab and iron hut on Sturt Creek, he delighted several of us until the small hours with his rendition of 'Pickwick's Trial', and other readings. At midnight, following our literary feast, we were served by the cook and partner, Weekes - a French scholar, no less - with white tea, yeast bread and butter. A veritable banquet for men accustomed to black tea, damper, and salt junk.

Mrs Stretch became the first woman to live on Sturt Creek which was more than seventy miles from any form of civilisation.

In later years Stretch made an unsuccessful bid to enter parliament for the Kimberley district of Western Australia. Frank Connor, of Connor, Doherty & Durack Limited, defeated him by one vote. Gordie felt that Stretch's philosophic temperament was reflected in his campaigning methods, and probably weakened his arguments against the more forceful speeches and personal appeals of his rival. Following his victory, when Connor the popular Irishman, stepped off the monthly mail boat on to the Wyndham jetty, Stretch was one of the first to greet him.

'Well,' said Connor, 'I suppose it's wigs on the green?'

'Yes, and drinks at the pub.' was Stretch's reply.

When Flora Valley was established, the owners had no money so business was based on barter. They operated like this for several years before opening a bank account. Gordie said that some of their first cheques circled the goldfields as currency before returning to them as payment for horse or bullock. Until paper orders known as 'shinplasters' were issued by the merchants, alluvial or leader gold was the main currency. Halls Creek storekeeper, Leopold Hirsohn charged five percent commission if a holder of these orders asked for gold, accredited cheques, or sterling.[9]

Most of the fossickers used gold as currency and lived a hand-to-mouth existence among the scorched, serrated spinifex-covered hills and quartz reefs. The ones that remained after the original rush were sometimes 'hatters' living alone in tents and bough sheds, searching the rugged ravines and gullies and scratching in old workings for a meagre living. Picturesque names were given to some of the prospecting localities: Flyweight Gully, Dead Finish Creek, and a particularly rough and almost inaccessible ravine was named 'The Gentle Annie' - Gordie suggests possibly from the old song, 'When the spring time comes, gentle Annie'.

It was only two or three years before the alluvial gold was worked out and most of the miners left the Kimberley for other more promising opportunities at Nullagine, Coolgardie and Kalgoorlie. Some reef mining continued for about ten years, but the big investors were attracted to more profitable sites and only a few of the old die-hard diggers remained. One was Jack Mantel and Gordie tells this story of him:

> Whenever he had a surplus, he handed it over to the only hotel bar in Halls Creek and drank till it was expended. I came to Jack's camp soon after his return from Halls Creek after one of these bouts. He greeted me with, 'Good job you didn't come yesterday, I'd have very likely shot you, I had 'em bad.'
>
> 'Had what?' I asked.
>
> 'The bloody d.ts. D'yer see that?' pointing to the mangled body of a frilled lizard. 'That's poor ol' 'Gladstone' He musta come and squatted on me bunk as he useter do, an' I musta thought he was an alligator.' We supplied Jack with Flora Valley salt beef at fourpence a pound about once in three or four weeks. It was fifteen miles each way along a rough bridle track, impassable for any vehicle. After his warning I was always careful to observe his camp from a distance before riding in.

Cattle killing by Aborigines was prevalent in East Kimberley. With the exception of a few horses being speared and run off in the early days, Flora Valley never had any trouble. Horses were twenty pounds per head then so replacement would have been costly. Gordie wished to avoid a repetition of the poor race relations on Wave Hill so when Big Jack, the headman of the 'Wongoo' tribe on Flora Valley asked if the tribe could come in and 'sit down longa station', he agreed. The only condition was that the cattle should be undisturbed. Once when Gordie shifted their camp off a central waterhole which was vital for the small herd, the Aborigines complained to the Halls Creek police. They had learned that the police were there for the protection of Aborigines as well as whites. When the Aborigines unwittingly speared a stray Flora Valley beast on Sturt Creek Station, Gordie received an unexpected apology. The excuse offered was that they thought the animal 'belonga Willie', referring to W. Stretch.

The Aborigines were allowed to remain on the property. They shifted their camp to the Elvire and the station provided 'killers' to feed them and punitive measures were never required. Through the still tropic nights the ancient, mystical sound of their corroborees drifted to the ears of the silent white audience lying in their swags on the verandah of the house.

The station had a number of capable and reliable Aboriginal stockmen, two of whom were mates, Daly and Darby. Once when there was trouble holding a mob of cattle on a dark and rainy night, Daly took his turn on night watch, circling the restless cattle while the rest of the men tried to get some shuteye. Each time he passed the sleeping camp everyone got the full benefit of his raucous tenor singing his favourite corroboree. Next day his mate Darby announced proudly, 'Daly singem goodfella; him gibbit lip hard fella.'

Another Aboriginal identity was named Gulya. He wasn't credited as much of a stockman, but he was intelligent and humorous, and a very adaptable and useful rouseabout. Jimmy Cullen enjoyed teasing him and that made Gulya a bit cheeky now and again. Dingo tails were worth five shillings each so Cullen used to set baits for them. On one particular occasion he warned Gulya to tie up his dingo pup because he was going to do some baiting. Gulya neglected to do this and his pup died from one of the baits. He approached Cullen in anger and distress over the situation. 'No matter, Gulya, plenty more longa bush. Where tail belonga your dead fella dog?' Gulya replied, 'Mindiegah', a vulgar and insulting expression. Surprised by Gulya's insolence, Cullen just laughed.

Jimmy Cullen had been managing Wave Hill when it was sold and he subsequently moved with the Gordons to Flora Valley. Gordie described Cullen as a hard man, but a good one. He had a quick temper accompanied with a rugged build and the fighting ability to back it up. Jim was a great horseman, and his wide experience enabled him to turn his hand to almost any task asked of him on a station. Although his quick temper sometimes made him hard on the Aborigines and animals and his expletives were unprintable, he was loyal, honest and absolutely reliable. Once his nose was broken by a kick from a horse he was shoeing. Using a pencil and handmirror, he paused briefly to set it back in shape before continuing with the shoeing job. Jimmy Cullen was tough.

Flora Valley employed a succession of horse breakers, two of whom were highly respected. They were Dick Smith and much later, George Kinivan. They could judge the spirit and temper of horses far better than most and used quieter, cooler methods. Slow, deliberate and fearless action was necessary to calm some of the bad tempered and nervous youngsters. Dick Smith could do almost anything with some of them. Gordie recalled:

> Once, after he had literally ridden into the bar of the Halls Creek hotel where he was a frequent and liberal patron, I came across him on his way home along the tortuous bridle track where single file was compulsory. He came along swaying happily in the saddle with his 'string' and blackboy behind him. As we pulled up he said, 'Now Mr. Buchanan, I'm teachin' thish colt carry drunken man, and when he falls to wait - hic - for'm. Now I'll show you. I'm s'pose to be drunken man, see, and I'm goin' fall off.' He did this very naturally. The colt stood calmly and let him slowly remount. 'Now see thish. I'll slide back over 'is rump and drop behind legsh.' He sat there on the ground for a few seconds stroking the colt's hind legs before he caught hold of the tail and pulled him-

self up to a standing position. 'Good tempered quiet colt thish, can't do it with most of 'em though!'

George Kinivan came to Flora Valley shortly before it was sold. He had gained his breaking experience in Queensland and had a way with equines. In about 1915 George became well known as a carrier in the East Kimberley region, using an enormous dray he had built himself from local timber, and a team of seventy two donkeys! Always keen on horse racing, he competed until well into his seventies, and rode many winners. No meeting was complete without George in the saddle.

Horses were the only form of transport and so it is understandable that the early pastoralists and adventurers were enthusiastic about horse racing. It wasn't long before meetings were conducted at both Halls Creek and Wyndham. These events were hotly contested and occasions for boisterous celebrations. Men often wandered across the north with a couple of ex-racehorses to contest the bush race meetings. The pastoralists were involved in the breeding of horses to use in the cattle camps and for transport, so it was quite natural for them to import a quality thoroughbred stallion to breed horses suited to the track. Nat, who always had a keen eye for a good horse, was responsible for breeding some quality gallopers on Wave Hill and later on Flora Valley.

The Kimberley Goldfields Jockey Club was formed in December 1887. The goldfields warden, J.M. Finnerty, was elected president and the two vice presidents were the Government Resident, Mr. C.D. Price, from Wyndham, and Nat Buchanan.[10] The annual Halls Creek races were held over two days and most of the crowd camped on the ground. The track was seven miles out beyond the rock-and-spinifex-bound township, on a Mitchell grass plain. Bough sheds were erected to provide shade under which the racing enthusiasts could gather.

A prominent figure around the goldfields at that time was W. Carr-Boyd. This big, hearty, amiable man had a loud booming voice, and a natural gift for yarn spinning. These attributes made him a brilliant story-teller, campfire entertainer and an amusing travelling companion. In the 1870's he wrote for the *Queenslander* under the pseudonym 'Pot Jostler', but some of his articles proved to be too risque for this paper and were rejected. Besides risking his life locating the remains of the Prout brothers, he was second in charge of Hodgkinson's North West Expedition in 1878. In 1883 he travelled with W.J. O'Donnell, who led an expedition from the Katherine Telegraph station to explore the country in the vicinity of the Ord River and Cambridge Gulf for the Cambridge Downs Pastoral Association.[11] Billy O'Donnell named the spectacular Carr Boyd Ranges after his companion.

W. Carr-Boyd's promising reports on the goldfields enticed many a would-be prospector to the Kimberley where most of them met with disappointment. Charging one pound per head, Carr-Boyd and O'Donnell piloted the first diggers from Wyndham to Halls Creek. O'Donnell was commissioned by the Western Australian Government to find a more direct route from Wyndham to the goldfields and shortened the original track by 140 miles. The O'Donnell Ranges and the main street of Wyndham were named in recognition of his early pioneering work.

In 1887, Gordie travelled with Carr-Boyd along the littered goldfields track and on the way passed many broke and disgruntled diggers quitting the fields. They were after Carr-Boyd's blood, claiming that he had misled them, so it was prudent to keep his identity a secret.

At one camp however, Carr-Boyd was unmasked and the situation turned very ugly. Somehow this incredible man with his disarming personality and ready wit calmed the angry men and re-established good humour. He then treated the diggers to an evening of unforgettable entertainment around the campfire. This man could talk his way out of anything.

Gordon Buchanan took up the Flora Valley leases in 1887 and after the sale of Wave Hill his uncles, Hugh and Wattie Gordon became partners in the property. However, all three Gordon brothers played an active part in the pastoral settlement of the Northern Territory and Western Australia. The first to quit the north was the eldest, Willie (W.G.). Since he first overlanded cattle to the Territory with Old Bluey in 1881-82, he had taken on contract droving. He overlanded what was probably the last mob of cattle to make the trek from Queensland to Western Australia for stocking purposes and suffered losses amounting to 400 head, from redwater fever. This was also one of the longest trips recorded - some 1,500 miles from Dalgonally to Lillamaloora in West Kimberley, with 1,000 head of heifers for the owner, Munro, one time Premier of Victoria. Lillamaloora was on the Leonard River about one hundred miles east of Derby.[12] With the proceeds of that and other successful droving trips, plus some profitable periods of station management, Willie had put aside a little nest egg for the future and retired to New England.

Hugh (J.H.M.) Gordon was born in New England in 1851 and was eight years younger than Willie. At twenty he managed Millie sheep station near Narrabri and later managed Ideraway Station in Queensland for a couple of years. Three years later he took his younger brother Wattie (W.R.) along when he drove 20,000 sheep from Craven Station to New South Wales. The two brothers joined Nat at Aramac in 1878 and were associated with him until their interest in Wave Hill terminated. From this time on the two bachelors never separated. Their attachment to each other was well known throughout the Kimberley. When a spinster governess came to live at Argyle she was warned that should she accept an unlikely offer of marriage from one of the brothers she would have to be prepared to take both of them! This comment, no doubt, caused the governess to act with caution.

Wattie had a reputation for being a great talker. Jack Watts, the mail contractor was a real livewire and a great friend of the Gordons and the Buchanans. One night at Halls Creek, Jack camped with his mate, Charlie Flinders, and their conversation turned to Flora Valley. Charlie said, 'I was talking to Wattie Gordon down there the other day.'

Jack quickly interrupted, 'You're a liar. I bet Wattie was talking to you!'

Hugh was the decision-maker and always had the final word, but not before Wattie provided all the adverse criticism. Wattie was fond of striking a pessimistic note, and what he had to say was minutely reminiscent and generally interesting - to strangers who hadn't heard it all before! When at the station, the three Flora Valley partners slept in their swags on the verandah of the homestead. Often, weary after returning from a mustering trip and wishing Wattie would shut up, Hugh and Gordie would pretend to go to sleep while he talked on. After a while he would say in disgust, 'Both asleep! What's the use of talking to you two. I can see it coming, we'll carry our swags out of this. But you won't listen. I'll turn in.'

The muffled response from Hugh's swag was always the same, 'Best thing you can do. We've heard all that before.'

Both brothers were outstanding horsemen and stockmen. Years later Jack Watts wrote to Gordie of the Gordon Brothers:

> How well I remember Hughie with the colts in the Flora Valley yards...and the night the cattle broke on Fox's Creek and I rode on old Wattie's left, hell for leather into the unknown. I would not like to say how I felt but old Wattie never turned a hair...and did you ever see him lag behind and those horses took some yarding...On the cutting out camp Australia will never see his equal.[13]

Flora Valley was a restful place for travellers and sometimes a hospital for them. The brothers and Kate were extremely generous with their time, energy and possessions. Charitable and community-minded, their material help extended to institutions and individuals throughout East Kimberley. On this wild frontier these sterling gentlemen set a fine example and were a great influence for good in the district.

In January 1892, shortly after Hugh finished his stint as manager of Wave Hill, he joined Gordie and Wattie in the Kimberley. With two Aborigines Hugh and Wattie set out on a cattle-hunting expedition in search of lost or strayed stock, and to check the waters on Sturt Creek. After watering at the Button they were encouraged by some scattered storms in that area, so continued east to Sturt Creek. It was an exceptionally dry year due to an almost non-existent wet season. The Aborigines attributed this to their local rain maker being in gaol for cattle stealing. All of the waterholes in Sturt Creek had dried up and at their last camp they found only a drop in a claypan from an isolated shower - barely enough for their outfit. The hardy pair pressed on to Wallamunga, seventy miles from Flora Valley, which they thought to be a permanent waterhole. They were astounded to find it dry. The usual noisy birdlife was absent and thirsty bush animals were dying in the glutinous mud, all that remained of a once brimming pool.

The heat was intense and their situation was very serious. It was too far back to the Button and sure water. Due to the terrific heat the claypan they had used on the outward journey would be dry before they could get back to it. They reasoned that the Sturt held more possibilities in its upper course of catching and holding water from local thunder showers, which were then threatening. So having just come fifty miles without water they decided to attempt to make the upper spring on the Swan, sixty miles away. There was a slender chance of survival if they went on but almost certain death if they went back. There would be no going back.

Jogging shirtless through the still and torrid night, accompanied only by the sounds of their progress and thoughts of their imminent demise, they reached the Sturt before daylight. They had to find water here because parched as they were, to make the trip across country to the Swan was suicide. A providential rainstorm had half-filled Wingramming Hole just below the junction of Sunbeam Creek and the reflection of the dying stars in the water was the most welcome sight on earth. They quenched their thirst and bathed their hot and tired bodies, the precious gift erasing the horrid spectre of madness and death.

Without that timely rainstorm it is extremely doubtful whether they could have pulled through. To have travelled a further thirty miles without water, in addition to the seventy five already accomplished, was an impossible feat in mid-summer. They completed the return journey aided by showers. Local Aborigines directed them to recent falls of rainwater but did not disclose the location of their own wells.

Following the loss of Wave Hill, Nat, Gordie and the Gordon brothers moved the cattle they acquired to Flora Valley. Even though Kate now lived at Flora Valley, Nat found that he could not rest there for any period of time. He was forever a rover, perhaps driven by circumstances as much as inclination. His name became a household word but he never had a permanent home until he was old and infirm.

Nat was once again employed by W.H. Osmand in 1895 when he took over the management of Ord River Station from Fred Booty. It was ten years since he delivered the first stock to that property. The Ord River cattle were notoriously wild. There were no brandings or musters for two years following delivery of the cattle because all Button's time was taken up deterring the Aborigines from killing the cattle, and packing supplies from Wyndham. Although he saved on cartage he neglected the essentials of branding and training necessary to establish a clean and quiet herd.

One day while Nat was watching operations on a cutting out camp, some cattle broke away. Although there were plenty of younger men to do the job, he instinctively galloped off to turn them back. The camp was on breakaway ground, innocent looking enough on the surface but a thin crust of earth disguised the cavities below. His horse stepped into one of these hidden holes and fell heavily.

Sitting with his back against a tree the old man fought back the waves of pain. Clouds of dust occasionally obscured the milling bellowing mob of the calf muster, and wild-eyed mickies and fizzing heifers shot out of the crowd and propped dead at the sight of him before running off. Hastie Burns galloped up and found his boss, 'You know Mr. Buchanan, you shouldn't go after them blasted old pikers like that. We can do it all.'

'Never mind Hastie,' said Nat 'catch that acrobat of mine. I'll ride the donkey after this - not so far to fall.'

Gordie never specified his father's injuries, but he did say that it took him a long time to recover and that the fall shook him considerably. Nat was in his sixty ninth year and old bones don't bounce. Although he would never admit it, he was in severe pain and had difficulty mounting his horse. Hastie escorted him back the few miles to the station. The head stockman had a great sense of humour and many quaint sayings which Nat was fond of drawing out.

'You'll get that mob all yarded before dark alright, Hastie?' he said as the stockman helped him to dismount.

'Too right, boss, we can do it on our heads.'

Nat chuckled delightedly at this response and said, 'Well, you know, Hastie, I tried that!'

Within twelve months Nat had resigned, the only manager except Leigh to avoid being sacked by Osmand who continually meddled in station affairs that he knew nothing about. This was more than a man of Nat's long experience could tolerate. Osmand was tight-fisted and reluctant to provide money for necessary improvements. He never visited the station in all the years that he owned it.

Nat wanted to explore the country west of Tennant Creek to Western Australia in the hope of finding a suitable stock route for the traffic of Barkly Tablelands and Queensland cattle to western markets. With the demise of her dream of a permanent home on Wave Hill, Kate, anticipating her husband's journey would be a long one, decided to return to Sydney for a time where she could assist in the care of her aged mother.

Although gold fever soon died, the pastoral industry went from strength to strength. Gordon Buchanan and the Gordon Brothers sold Flora Valley to Vesteys in 1914.

In his book *Gather No Moss* written in conjunction with Lynda Tapp, Billy Linklater, well known bushman and poet of the times wrote:

> ...the Gordon's were given a send off that was the talk of the Kimberleys. The assembled stockboys with their lubras and children cried and cut their heads till the blood streamed down their faces; among the blacks the letting of blood must accompany the expression of really poignant grief. The Aborigines express their belief that the heart is the seat of the emotions by saying, when deeply moved, 'I bin cry in my binji.'

Flora Valley is still a flourishing pastoral enterprise presently owned by the Heytesbury Pastoral Group. Some relics of the old buildings survived until the late 1980's when what hadn't decayed was unfortunately knocked down and a new homestead built.

References

1. Darwin Shipping and Passenger lists, NT Archives.
2. G.C.Bolton. 1953. *A Survey of the Kimberley Pastoral Industry from 1885 to the Present.*
3. Police Occurrence Books, Kimberley-Elvire Creek 1886-1887 and Halls Creek 1887-1888. AN5/Halls Creek, Acc. No. 1422, W.A. Archives
4. Police Occurrence Books, op cit. Entry dated 23/101886.
5. Wyndham Police Occurrence Book, entry dated 18/12/1887, 30/7/1891, AN5 Wyndham, Acc. No. 741, W.A. Archives
6. Letter from P.B. Watts to G. Buchanan, dated 10/10/37. Buchanan Collection.
7. Letter from P.B. Watts to G. Buchanan, dated 2/3/37. Buchanan Collection.
8. Letter from J.W. Durack to G. Buchanan dated, 25/11/33. Buchanan Collection.
9. Hirsohn arrived in Halls Creek on 9/11/1887, according to the entry in the Halls Creek Police Occurrence of the same date.
10. G.H. Lamond, 1986. *Tales of the Overland.* Appendix 4. Hesperian Press.
11. E. Favenc. 1888.*The History of Australian Exploration 1788 - 1888.* Turner & Henderson. Syd. 1888.
12. *The Northern Territory Times & Gazette,* 2/2/1889. *Sydney Stock and Station Journal,* 19/5/22.
13. Letter from P.B. Watts to G. Buchanan, dated 3/3/36, 15/10/36, 2/3/37.

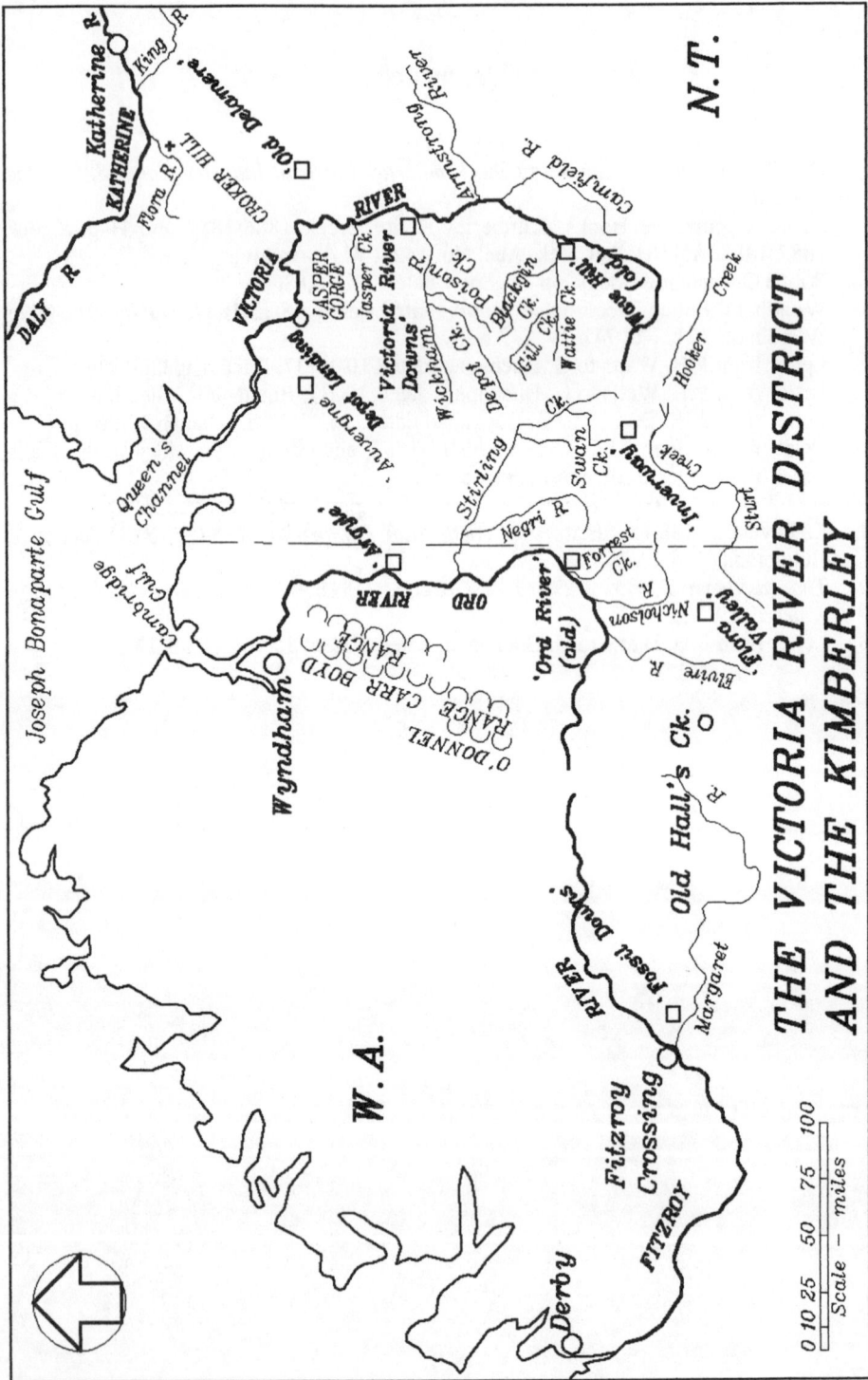

The Victoria River District and the Kimberley

31. Chain Gang of Aborigines convicted of cattle killing - Wyndham, W.A.
Buchanan Collection

32. Aborigines cutting up a "killer" outside the yards at Flora Valley Station
Buchanan Collection

33. Supplies being delivered by horse team to Flora Valley Station
Buchanan Collection

34. Aboriginal staff and outbuildings on Flora Valley Station
Buchanan Collection

35. Standing from Left - Hugh Gordon, Wattie Gordon
Seated from Left - (?) Jimmy Cullen, Gordie Buchanan, (?) Willie Glass
Buchanan Collection

36. "Daly" on Flora Valley
Buchanan Collection

17.

Crossing the Janami

1896

———❦———

To prevent the spread of cattle tick into the Kimberley from the Northern Territory an embargo had been placed on cattle entering Western Australia. This was eventually lifted, although the stock tax still remained. Anticipating the eventual lifting of the embargo Nat decided to explore a more direct route for stock crossing the Barkly from Queensland and the Territory, through Sturt Creek in the West. If a suitable route could be found it would also be advantageous to pastoralists on the Tablelands and the McDonnell Ranges. Armed with letters of introduction Nat approached the South Australian Government for support for the exploration.[1]

In January 1896, the Northern Territory Government Resident, C.J. Dashwood, telegraphed the Hon. C.C. Kingston, premier of South Australia, advocating that the exploration be undertaken. Nat was highly recommended as the most experienced and capable man available to undertake the job and Dashwood suggested to Kingston that the loan of camels for the expedition would be a useful contribution.[2]

Nat Buchanan, whom the Government have chosen to search for a new stockroute to the westward, is an unassuming explorer who has a far better claim to renown than most of the crowd who have posed of later years as Australian explorers. He must be now seventy years of age and he is possibly too old to stand many severe bush crises, but in his best day nothing was too rough for him. He has a wonderful instinct for direction in the bush, and could travel straight as a dart to a given point. Moreover he can go almost as long without water as a camel...Hundreds of bush yarns are current about 'Old Bluey', as stockmen familiarly call him, but in none of them have we ever traced anything but the most flattering recognition of the good work done by Buchanan in the pioneering days of the Far North.

The Northern Territory Times & Gazette, 7/2/1896.

The South Australian Government agreed to loan Nat six bull camels valued at 240 pounds, and equipment to the value of twenty five pounds, for a period of eighteen months. The conditions were that he took every care of the camels and equipment and used them only for the purpose of searching for a stockroute from the Overland Telegraph line to Lake Wells or Mount Margaret in Western Australia. The camels were to be returned upon the expiry of the designated loan period and the cost of any stock or equipment lost was Nat's responsibility.[3]

Nat recognised the camels' special advantages in desert country and would not have undertaken this trip without them, but he was unfamiliar with the animals and refused to ride one. At Oodnadatta he added the six camels to his plant of six horses and took on a man to help him as far as Tennant Creek. Here he anticipated recruiting a couple of reliable men willing to undertake the crossing.

The two men made their way north along the Overland Telegraph line and at Barrow Creek one man was engaged to make the crossing. At Tennant Creek, despite telegraphing messages up and down the line he was unable to find another person willing to undertake the next leg of the trip. This resulted in the man he had hired pulling out. With six horses and a string of camels it would have been impossible to attempt such a hazardous venture alone. Nat was becoming restless waiting around so the telegraph operator recommended an Aborigine in his employ. Jack was a tall local tribesman who could speak a bit of pidgin English, had some experience with camels, and was familiar with the country for about seventy miles west of Tennant Creek. Nat promptly came to an arrangement with Jack, who agreed to make the crossing.

Impatient to set out, all attempts to persuade him to wait for more help fell on deaf ears. The desert or the bush held no terrors for this experienced pathfinder. With 'Camel Jack', as his sole companion he set out on 24th April 1896. The South Australian Register published the brief daily record kept by Nat on this, his final and possibly most remarkable journey.[4] Because this appears to be the only personally written account of any of his explorations it seems appropriate to include it.

Tennant's Creek, April 24, 1896 - Not being able to get men here I decided to start without any, taking only one black boy kindly lent me by Mr. Besley. At Barrow Creek I did engage a man at fifty shillings per week and a seven pound guarantee for each of his horses, but here, as I could not get another to engage, he refused to come any further with so small a party. Travelled north north west ten miles, and camped on Bishop's Creek, running north east ten miles.

25th - Steered north west by north, and camped on a small creek. Country today and yesterday similar to Tennant's Creek; fourteen miles.

26th - Shaped a north west course, and in a few miles reached a sandstone tableland, and camped at a blacks' well; thirteen miles.

27th - Travelled west north west to a creek with rocky waterholes, which should last for three or four months. Country sand and scrub; the creek losing itself in sand flats; sixteen miles.

28th - For nine miles north west passed over stony country, with low scrub, bloodwood, and desert gum, and struck a creek trending west, with small rockholes; nine miles.

29th - continued north west course five miles, when the boy pulled up at a small waterhole, saying the next was too far away to reach that day; five miles.

30th - resumed same course today, and travelling along foot of low sandstone ridges lying to the north east reached water in a small creek flowing out of them; thirteen miles.

May 1st - Thirteen miles again north west brought us to a big swamp or lake in a coolibah or box forest. The water at the edge was shallow, but the black boy said it was permanent and from ten to twenty miles long. The boxtrees grew out in the water as far as I could see; thirteen miles. The blackboy, Jack, who, since we left Tennant's Creek has been leading me to the waters that he knows of, has been otherwise almost useless to me, as he refuses to do what I tell him, and I have to pack up and do everything else myself. However, I was willing to put up with this so long as he continued to ride his camel in the lead on a westerly course, as I rode a horse, and could keep the rest up behind him. But from this point he wanted to go north north east, no doubt expecting to meet some of his tribe in that direction, and was quite master of the situation, as I could not afford to let him go, being unable to manage both horses and camels myself. Before turning in, I thought of a plan, and while he slept I secured him by his leg with handcuffs to his camel riding saddle at the end of a chain about three feet long, and thus secured him where he would be most useful to me as he could put on his own saddle and help me pack the water casks without much difficulty; thirteen miles.

Nat's plan to capture the Aborigine depended upon his deftness and stealth, and the depth of Jack's slumbers. The night was mild and Jack slept on his blanket beside the camel pack and riding saddle. The chain was first attached to the riding saddle so that if he should wake and be suspicious, Nat could use the excuse that he was just altering the gear. While Jack slumbered, the handcuff was carefully attached to his ankle over a corner of blanket.[5] Mission accomplished, Nat woke Jack and attempted to explain to the reasons for his captivity. Jack didn't appear to comprehend the explanation and was understandably very sulky. Perhaps he felt that his situation was even more desperate than his gaolers. Jack usually managed the camels but with his movements now limited Nat's workload was greater as he had to muster both horses and camels each day.

> *2nd - With boy in lead we skirted the lake for five miles, then leaving it in seven miles reached a small well where I could get no water without help, but as good camel feed was plentiful camped here; twelve miles.*

> *3rd - Resumed course to another well and got water for the horses only, but there is good camel feed; thirteen miles.*

> *4th - A shower of rain fell this morning. Started in the evening north west by north, and in seven miles reached a lake or swamp into which I could see no distance for timber; seven miles.*

> *5th - Five miles on same course brought us to a peculiar rocky well with plenty of water. As this is the last water known to Jack made preparations for dry stage. Fair camel feed here; five miles.*

> *6th - Started west by north over comparatively open country, which is apparently a high tableland and camped without water in thick scrub; fifteen miles.*

> *7th - Continued same course and camped at rock well, similar to the last, not quite so large, but having plenty of water. Good camel bush here. A heavy black smoke hung ahead of us all day, while there were fires behind us and at night fires in all directions. Both these last rockholes I take to be permanent. Another close to this camp was half filled with mud; fourteen miles."*

The smoke referred to was the result of fires in spinifex and turpentine bush, generally found together in that dry inland country.

All this time and later on, Nat never referred to the hardships, isolation or potential danger from Aborigines. Not since he had to chain Jack did he do more than record the distances and direction of travel, and the few occurrences of water or other salient features on his daily course. Gordie commented:

> *He took a matter of fact view of those things, perhaps a fatalistic view, and neither mother nor I could draw much out of him. He had been long accustomed to blacks, bushfires, waterless regions and loneliness.*

Although unaware that he had just started out on a long dry stage, Nat was nevertheless prepared for any eventuality.

> *8th - Steered west through high desert country and camped in heavy spinifex. No grass or water; twenty miles.*

> *9th - Started west through similar country to that passed yesterday. In evening saw ranges to west south west; twenty eight miles.*

> *10th - Shaped a course west south west for ranges seen yesterday. This afternoon the smoke from surrounding bushfires became so dense that*

I could not see the mountains or anything at a distance. Camped, no water. We passed a small blacks' well in the afternoon showing at the bottom a little water and a coolamon lately left by the natives, whose tracks were plentiful here. Could have watered camels and horses here with some help, but I could not let the boy loose so near his own country, and as I was not able for the work of deepening and widening the well continued my course, hoping to get a better supply further on. Camped, no water; twenty one miles.

11th - Determined this morning to make for sure water at the head of the Victoria River. Eight miles west north west, the smoke having partly cleared away, could see mountains to the south west about twenty miles, and about ten miles in the same direction three conical or peaked hills having the appearance of auriferous country. Continued course ten miles and camped without water.

These hills were subsequently named Buchanan Hills.

12th - 13th - Yesterday the country was fairly grassed and not much spinifex, with patches of limestone and jasper; today mostly desert and one mile of plain brought us to water in a basalt creek running through the plain which extended to the south south west to the horizon; four miles.

Water at last! The camels were truly remarkable because despite carrying casks of water for the horses and travelling for six days without water themselves, they were not unduly distressed.

14th - Camped today for a rest.

Jack had been confined for two weeks so having arrived at water and being on the outskirts of country he knew, Nat decided to liberate him. The old man was exhausted and needed the assistance of the Aborigine to help with the camels and horses but considered Jack's welfare ahead of his own. Without the Aborigine's help he would have had to release the camels and then reimburse the South Australian Government for their loss. When Nat offered Jack a waterbag, tucker and tobacco with which to return to his own country, Jack refused to go, saying, 'Me stop longa you, mind 'em camel.' Very well he did it too, and all the packing and other jobs which he had previously refused. This eased the burden considerably for Nat who now only had the cooking and the horsehunting to attend. It was a great relief knowing that now he could hold the camels.

15th - Ran the creek up to top water west south west and camped; four miles.

16th - Continued west south west course for six miles and camped. No water; six miles.

17th - Struck Hooker's Creek today, the bed of the creek being dry and sandy and apparently not run this year. Found water in a billabong, which was nearly dry. Course today west south west; thirteen miles.

18th - After two miles due west over good country entered a forest fairly grassed and camped without water; west, eighteen miles.

19th - Early this morning passed a sandy creek with granite banks, coming from the west south west and nine miles crossed another similar, but having water in it. In an hour's further travelling reached a granite range, and made a dry camp on a high tableland three miles beyond the range, which forms its eastern edge. Country passed today well grassed having patches of spinifex in places. Course west; fifteen miles.

20th - Started south west by west to get a view to the south of some of the heights around. In ten miles saw five stray horses, which must be watering somewhere near. Six miles further we saw a valley and open plains to the south west, and in three hours camped on the western limit of the tableland, about one hundred feet high, south west by west; twenty six miles.

21st - Reached the southern bend of Sturt's Creek after five hours travelling (south west by west) over alternate forest and spinifex country of basalt formation, mostly fairly grassed; fifteen miles.

22nd - Steered west south west over well grassed open plains and forest country and camped without water; seventeen mile.

23rd - Starting on a west north west course, in seven miles changed to north west, and in eight miles struck Sturt's Creek again, about the border of Western Australia.

24th - one camel died here last night. The country from here to Halls Creek is well known, and the distance about ninety miles.

With the remaining five camels and the horses, Nat and Camel Jack arrived at Flora Valley in the first week of June.[6] Another unexplored region of Australia had been successfully negotiated by this superlative bushman. Although there were patches of good feed along the way, the lack of reliable surface water prohibited a stockroute from Tennant Creek to Western Australia by the course Nat had taken. This courageous crossing of the desert country brought no financial gain but added considerably to the legend of this remarkable pathfinder. Of course he was still obliged to pay forty pounds for the camel that died! Jack remained on Flora Valley tending the remaining camels until after the wet season when, with the South Australian Government's blessing, they were handed over to David Carnegie.[7]

Nat never fully recovered from the fall he had on Ord River and he undertook this last great journey against the advice of his doctor. Although he chose to ignore the signs, advancing years and deteriorating health were catching up with him. He made several more desert journeys after this and Gordie felt that the hardships suffered on these shortened his life.

References

1. *Letters of introduction*, Aust. Archives, A.C.T. A1640 96/35
2. Telegram from N.T. Govt. Resident C.J. Dashwood, to Hon. C.C. Kingston, Premier of S.A. dated 20/1/96, Aust. Archives, A.C.T., A1640 96/35.
3. Loan Agreement, Aust. Archives, A.C.T. A1640 96/35.
4. S.A. Register 10/10/96, S.A. Records Office GRS 9/5
5. Nat presumably carried a pair of handcuffs with him because previous experience dictated that these were an essential piece of equipment on long trips into unknown country.
6. Halls Creek Police Record Book, entry dated 6/6/96, W.A. Archives, Acc No. 1422
7. Telegram from Nat Buchanan to F. W. Holden, dated 26/3/1896, Aust. Archives, A.C.T. A1640 96/35.

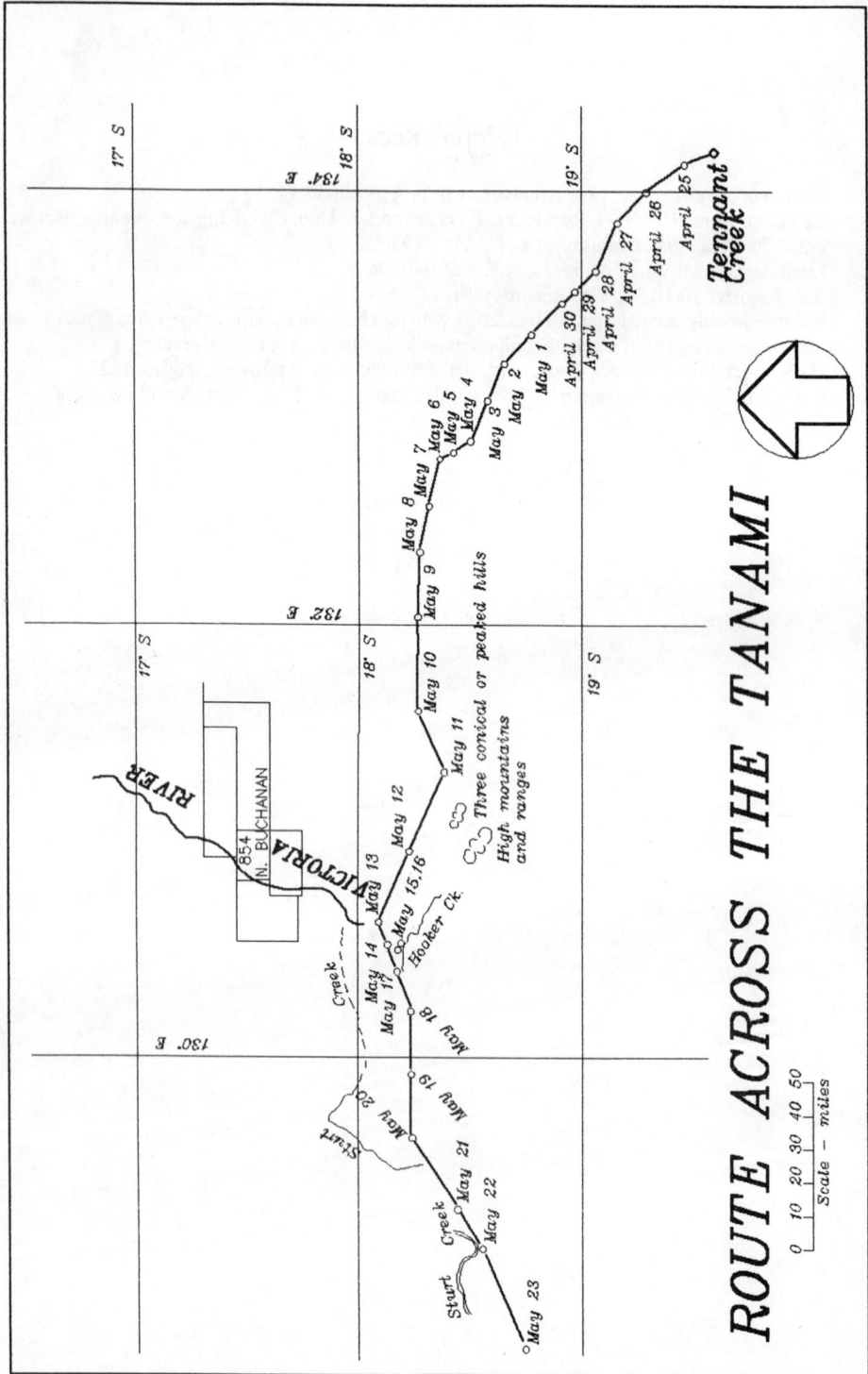

ROUTE ACROSS THE TANAMI

Route across the Tanami

37. A camel ready for the road
Buchanan Collection

38. Rockhole on the road to the Tanami
Buchanan Collection

18.

Farewell to Exploring

1896 - 1899

—⁓⁓⁓∞⁓⁓—

After resting briefly at Flora Valley, Nat, wanting to get the best use out of the camels before they were returned, continued explorations into the desert in search of that elusive stockroute to Western Australia. With Camel Jack he revisited the country they had recently crossed south of Wave Hill station, probably to reassess the availability of surface water, and it appears that he felt that he had succeeded in his mission.[1] The fact that the next step of his exploration, accompanied by Jimmy Cullen and Jack, was to search for water between Sturt Creek and the Oakover River in the direction of Joanna Springs indicates that he was pushing on with his original objective. When he failed to find sufficient surface water in this area he still refused to give up, deciding instead to attempt a passage through to the Oakover River from the Fitzroy River.

In this same year an expedition sponsored by Albert Frederick Calvert, a wealthy mining engineer, set out to find a stockroute by which cattle could be walked south from the north of Western Australia, thereby saving shipping costs. The leader of this party was Lawrence Wells. He was accompanied by his cousin Charles Wells, George Jones, two other men and two Afghan camel men, Said

Ameir and Bejah Dervish. At Separation Well the party broke into two groups to better explore the country. The arrangement was for the two parties to reunite in two weeks at Joanna Springs. The contingency plan was to continue on to the Fitzroy if either party failed to arrive. Both parties were unaware that they were following a map made by Warburton in 1872, where he had inaccurately mapped Joanna Springs. To cut a long and tragic story short, Charles Wells and George Jones became lost and perished. Search parties had been organised to find them and Larry Wells was about to make his second attempt to search the country between Fitzroy telegraph station and Joanna Springs. Here he met Nat, complete with camels, and about to make an attempt to find a route to the Oakover by tracking south from Fitzroy telegraph station.[2]

Nat offered his assistance in the search for the missing men and Larry Wells was relieved to have a man of his experience along. With Bejah Dervish and George the tracker, they set out on the 4th December 1896. Between them they had eight camels, but the four belonging to Wells were in poor condition. They travelled via GoGo Station, Noonkanbah and Kallaida Stations into arid country to a spot past Wells' previous point of return. By following the tracks of a Myall Aborigine they came to a small rockhole which the camels soon drank dry. After dark the thirsty Aborigine was caught while attempting to creep into the dry rockhole. Wells gave him food and water and he then willingly directed them in a south west direction to the next water.

That day the party came across a fifteen year old, lone Aboriginal boy camped in a wurly near a soak. He was unafraid and it appeared that his isolation was due to some tribal custom. By baling the soak with a quart pot for the best part of a very hot day, the men managed to deliver three buckets of water for each of the camels. The Aborigines called the spring 'Ngowallarra' and Wells named it 'Welcome Water'. Next day their guides directed them to another soak but after digging the drift sand in the terrible heat, it produced insufficient water for their needs.

Nat's camels had deserted during the night so an early start the following morning was aborted. By the time the camels were found it was so hot they had to wait until late afternoon before setting out. While the party was resting the two Aborigines tried to leave the camp. The older of the two was successful, but the boy was captured and then restrained from further escape attempts by a three foot leg chain securing him to a camel saddle. Because feed was scarce and this was the hottest part of the year the camels needed water every three days. Nat and Wells were acutely aware of their dependence on the Aboriginal boy to guide them.

That night when they set out the boy rode with Bejah on his camel and directed the party to two soaks in close proximity to each other, called by the Aboriginal name 'Kullga-ngunngunn'. In the morning they saw that the soaks were about 30 feet in diameter and had been excavated by the Aborigines to about ten feet deep, providing plenty of water for their needs. The soaks were surrounded by rushes and acacias and contained a few inches of clear water over a sandy bottom. Wells' camels developed symptoms of Gastrolobium poisoning here and despite drenching with Epsom salts one died, causing further delay.

The boy's chain was released from the camel saddle so he had the freedom of the camp, but remained attached to his ankle. This proved to be his undoing because when he again tried to escape it snared in a bush, enabling Wells to catch him. Fearful of retribution, the boy threw himself upon George and Bejah but was pleasantly surprised when following his restraint to the camel saddle, he was given food and water. Further delay was caused when one of Nat's camels sickened and died after failing to respond to treatment. Having lost two valuable camels at this soak Wells no doubt felt justified in naming it 'Disaster Water.'

Christmas Day was not a festive occasion. The party broke camp early but after four hours travelling it became so hot that they were forced to camp again. The following day the boy appeared to be confused about the locality of the next soak and gave conflicting directions to the same water. The men thought that he might be withholding information and to test this they deprived him of water for ten miles in the hope that thirst would prompt him to give the correct direction. When they attempted to proceed further south the boy became panic-stricken and the men realised then that he was beyond his tribal boundaries and genuinely could not help them. They gave him a good drink of water and then pitched camp.

In the evening of December 27th the sound of thunder and the appearance of storm clouds were welcome. The water kegs were dry and each man only had half a pint left. Before settling their parched, exhausted bodies down to sleep, a hole was dug in the sand and a tarpaulin arranged over it in the hope of catching some rainwater from the impending thunderstorm. The tarpaulin caught enough water to provide each man with a small drink. Ninety miles out from the Fitzroy River, Wells and Nat found themselves in desperate circumstances. With their water nearly gone, the camels sick and exhausted and precious little available camel feed, the only sensible course was to turn back. The Aboriginal boy was freed and to save the camels the party walked much of the way back to Fitzroy Crossing. The remains of the perished men were discovered by another search party, quite close to Joanna Springs.[3] If nothing else, this harrowing journey proved to Nat the improbability of finding a well-watered route to the Oakover River from the Fitzroy.

On his return to Halls Creek, Nat met the Hon. David Carnegie who had been exploring the western desert from south to north. With the permission of the South Australian Government the remaining borrowed camels were handed over to him.

Carnegie was given a civic reception at Halls Creek which was attended by twenty or more residents and visitors.

> *The reception was held in the only hotel in the township and was presided over by the Warden and Magistrate, Mr. Cummins, a*

scholarly man who had lately become engaged to be married. The banquet was held within the whitewashed mud walls of the long dining room. The table was flanked by long forms, or stools, with a short one at each end. There was no other furniture in the room except for two or three beer cases which acted as sideboards. A coatless and careless assembly elbowed each other round the oilcloth covered table. Initially many of the guests were diffident and quiet, but after a few drinks they loosened up.

Upon hilarious calls from the company, W.H. Stretch rose to his imposing height to make the first speech. He began, 'This is an age of exploration. Nansen is exploring the North Pole, and our friends Carnegie and Buchanan are exploring the deserts and semi-deserts of our Australian inland, and our worthy friend and worshipful Warden is to explore the happy climes of matrimony.' There was an enamelled dish of tomatoes at the other end of the table, and Cummins, in the merry mood of the moment, selected a ripe one and let Stretch have it full on his shirt front.

Nat's fascination with the uncharted desert country never waned and he continued to make exploratory forays in an attempt to discover the secrets it held. Shortly after his return from the search for Wells and Jones, he set out with his nephew, Harry Farquharson, about 100 miles into the arid interior south-east of Sturt Creek. There was no pastoral country in that region but they did discover some low grade mica deposits. Without camels further penetration of that almost waterless country was hopeless.

After the wet of '96-'97 Nat set out with Camel Jack and George on a prospecting tour to Buchanan Hills. His purpose was twofold, wanting to return Jack to his own country as well as investigate more closely those strange conical hills which he suspected might contain gold. Both objects fulfilled 'his inherent desire to be on the move into some faraway region beyond the restraints of civilisation'. Although they had twelve to fifteen horses with them Jack refused to ride. If he couldn't ride a camel he preferred to walk. He was not unduly disadvantaged as his long stride could match the walking pace of almost any horse.

Upon closer investigation the solitary Buchanan Hills, which had captured Nat's imagination on his westward journey because they appeared to be gold-bearing in nature, revealed no treasure. However, Jack, now near his own country, was provided with a waterbag, tucker, tobacco and a tomahawk and set out happily for Tennant Creek where he arrived safely. Before he left Flora Valley the station Aborigines had given Jack two spears and a woomera as a parting gift. A telegram was sent in advance from Halls Creek to Tennant Creek advising that Jack should be reprovisioned on his arrival and the cost charged to Nat. Although Jack was initially an unwilling and uncooperative servant, from the time he was released from the camel chain he had stuck faithfully by the old man and served him well.

This was a time of parting. First Camel Jack, then George, that incomparable tracker who had been a trusted and devoted servant for about fifteen years, was returned to Wave Hill. Nat's relationship with the Aborigines who worked for him was a very special one built on trust and mutual respect. It was said that, 'Buchanan had the best blacks in the Territory.' He was a quiet, even-tempered man who was true to his word and would not ask of others what he was not prepared to do himself. Aboriginal stockmen were known to state with pride, 'I bin stockboy longa Nat Buchanan.'

From Wave Hill to Flora Valley the Aborigines referred to him as 'Paraway'. The name came about because whenever they asked, 'Which way you walk now, Boss?'

Nat's inevitable reply was, 'Long way, faraway'.

Even a race of people who were essentially nomadic found Nat's relentless journeys incomprehensible. They would regularly ask him, 'What for you no sit down quiet fella longa Plora Valley?'

His answer was always the same, 'No good sit down, me walkabout.'

Finally, Nat said goodbye to the arid country of Australia's interior. Afflicted with a bad heart and asthma, he defied his doctor's orders to take one last journey across the vast country he had been instrumental in opening up to settlement. The purpose of this final mission was to buy Queensland horses and overland them to the Kimberley but this was probably an excuse to say goodbye to the country he loved. M.P. Durack recalled his last meeting with Nat at Camooweal, towards the end of 1898. Nat, who was on his way to visit Donald McIntyre at Dalgonally, was camped with an Aboriginal travelling companion on the Georgina. The Camooweal race meeting was in full swing, and hearing that M.P. was in town, Nat sought him out. The two pioneers talked long into the night. Nat was in a reminiscent mood and dropping his customary reserve, related to M.P. the story of his Tanami trip.[4]

With the help of Charlie Sweeney, plus an Aborigine and his woman he collected the horses from an unknown source and set out along the Old Gulf Track. The usual route across the Barkly was much too dry to attempt that year. In order to take a more direct line west via Newcastle Waters and the Murranji Track, Nat left the Gulf Track at Borroloola, and pushed south west via Laila Creek and O.T. Station through the northern Barkly, until he reached Newcastle Waters. The route he used was eventually opened up to horse-drawn vehicles when the replacement of telegraph poles and the duplication of the Overland Telegraph Line took place. The Government found that it was more economical to ship the equipment needed for the renovations to Borroloola, and then transport it by team to Powell Creek, than to overland it directly from Darwin. Bill Weldon was pilot for the first teams to make this trip.[5]

Twelve years after blazing the Murranji track and at the age of seventy four, Nat had again made a successful crossing. The 100 horses were delivered to their

Kimberley destination in early 1899. Truly an extraordinary feat by a remarkable man.

By the time he retired from the north, the bush was buzzing with stories of his incomparable bushcraft, his phenomenal sense of direction and his ability to survive in waterless inhospitable country. Although he was reserved and modest he managed to capture both the admiration and the imagination of the bush people. They repeated stories of their own encounters with him and those of others, thus the legend of Nat Buchanan continued to grow long after he was gone. Because he was called 'Old Bluey' and carried an umbrella as a sunshade, the new storytellers who took up the tale after those that knew him personally had passed away, made the incorrect assumption that he had red hair. Today the popular image of Nat is that of a tall, bearded, redheaded man mounted on a camel and sporting a green umbrella. His use of camels as pack animals during the crossing of the Tanami and subsequent desert forays has led to the misconception that Nat rode a camel. This was never the case, he always rode a horse.

Nat's 'bump of locality' was the subject of many of the stories that circulated about him. As a young man, C.E. Gaunt crossed the Barkly Tablelands with Nat. He recalled that when they were travelling from Eva Downs to Powell Creek, they stopped for a spell on the open downs country between Bundara and Monmona Creeks. After resuming their journey Gaunt discovered that he had left his favourite knife at the resting spot. Nat assured him that they would pick it up on the return trip, but because of the absence of landmarks on the downs, Gaunt was sceptical. On their return Nat was true to his word. Although he did not follow their outward tracks he still rode directly to the location of the lost knife. His sense of direction was uncanny.[6]

Jimmy Cullen was fond of relating an incident which occurred when he and Nat were travelling together. The pair, with three spare horses and a packhorse, set out for a rockhole in the semi-desert country about sixty miles south east of Wave Hill. Nat rode in the lead. They rested for a few hours at noon and then went on until sundown, rested again, and then proceeded in darkness. Cullen drove the horses along behind through the dark and featureless countryside. Never hesitating, Nat went on, as straight as a dart, and in the early hours of the morning he stopped and lit a match to consult his fob watch. He told Cullen that they would be at the rockhole in another hour and a half. True to his word, and in the given time they arrived at the brimming rockhole. Although an experienced bushman, Cullen was awed by this display of bushmanship. Nat did not follow any creek and had no landmark to go by. He just set a direct course from the station and without deviating from it, rode straight to the rockhole.[7]

Nat travelled so widely that there were few people in the bush who didn't know him by sight. Tommy Cahill was fond of recalling the time Nat managed to pass through Wave Hill, unrecognised. Travelling with a string of pack horses and an Aborigine on one of his many unrecorded trips into the arid country around

Hookers Creek, he called at the homestead. The station was almost deserted because everyone, with the exception of the new-chum bookkeeper, was out mustering. Nat made a distinctive figure riding through the outback under the shade of his umbrella and was accustomed to being recognised, so without introducing himself he passed the time of day with the man and casually mentioned that he was heading south to the desert country. The bookkeeper became alarmed and tried very hard to talk him out of it. Despite the well-intentioned warning a bemused Nat rode on. When Tommy Cahill returned the bookkeeper reported the incident, describing Nat as an 'old city josser, riding with an umbrella and some packhorses in tow', who refused to listen to reason. Tommy had no trouble recognising his old boss from this description and burst into peals of laughter. He saw the funny side of the 'Master Mariner of the Desert' being warned of its dangers by a new chum bookkeeper!

> Buchanan had the gift of bushmanship, location, observation, and initiative.
>
> He had great organising ability and he knew how to handle men.
>
> Quiet and unassuming, but with incredible will power, he was a genial comrade and one who would always stand by you in time of need.
>
> The Northern Standard, 19/6/1934.

References

1. *Northern Territory Times & Gazette*, 4/12/1896
2. *The Proceedings of the Royal Geographical Society of Australasia*, S.A. Branch, Vol. 2 & 3, P.2161, Sessions 86-87, 87-88, Mortlock Library, Adel. S.A. and W. & C Steele. 1978. *To the Great Gulf -Surveys and Explorations of L.A. Wells*. Lynton Publications Pty Ltd.
3. Telegram dated 26/3/1897, Aust. Archives A.C.T. A1640 96/35.
4. Mary Durack. 1983. *Sons in the Saddle*. Corgi, 1985.
5. W. Linklater and L. Tapp. 1968. *Gather No Moss*. Macmillan.
6. *The Northern Standard*, 19/6/34.
7. *The Northern Standard*, 19/6/34.

39. Desert Aborigines
Buchanan Collection

40. Mr Thomas Cahill - head stockman and relieving manager of Wave Hill
for Nat Buchanan and manager for W.F. Buchanan
Sydney Morning Herald, 3/2/22

19.

Homecamp

1899 ~ 1901

━━━━━━⟿⧜⟻━━━━━━

Nat's age and infirmity drove him from the work and country he loved. He bought 'Kenmuir', a small lucerne farm near Dungowan in New England. The farm was situated on Dungowan creek which wound between the ranges and spurs of the great New England plateau. The area was settled mostly by dairy, maize and lucerne farmers but on the steep mountain sides the larger properties ran sheep. This beautiful spot became Nat and Catherine's final home.

Nat often drove the twenty one miles into Tamworth in a very light, one horse buggy. He soon got to know many of the settlers and was often delayed talking to them until after dark. This made Kate anxious for his safety and not without reason. The buggy had no brakes, the harness was in need of repair and the horse had the reputation of being a bolter. This was more likely though when he saw the winkers coming. Once when about to leave Tamworth for home Kate noticed that the breeching strap was partly worn through and chided Nat about it.

'Look at that strap, Nat. It's dangerous and might break at any time. Better get it mended.'

Surveying the outfit with his usual calm he replied, 'There's the tail Kate, there's always the tail,' indicating that the crupper would take the strain.

He was generally careless about his dress in the country and frequently Kate had to arrange his apparel for him. With the exceptions of when he was on the road or visiting neighbours, Nat was a slow starter, and Kate had to get him organised. Once, when they were going to Tamworth, she was too busy to attend to her husband but asked him to be ready by a certain time. Kate appeared immaculately attired at the appointed time to find the buggy ready at the gate, and the horse stamping his feet and shaking his head at the flies, but no sign of her husband. 'Whatever can be keeping Nat?' she enquired. Much to Gordie's amusement his father appeared half clothed at the bedroom door and bawled, 'What are you going to put on me Kate?'

Despite his age and poor health Nat worked the 20 acre lucerne patch until the end. Although he knew he might collapse any day, to quote Gordie, 'Like Tolstoy's peasant, he went on ploughing.'

During his life the most devastating blow that befell him was the loss of Wave Hill. Ironically, his failure to reach affluence may have assured his success as a drover and explorer. Perhaps riches would have restrained his adventurous spirit and restricted his activities to easier and more profitable pastoral paths.

> *The urge to explore and discover whatever the heart of Australia held seemed to be his ruling passion. In most of the adventures which made him famous it was his misfortune to be handicapped either by too little equipment or too much. In some cases he had only a blackboy to rely on, and in others he had to pilot mobs of cattle. By the first method he put his name on the map and by the second he led the van of pastoral progress in the north.*

When the end came in September of 1901, he was buried on Kenmuir and his son planted Kurrajong trees to mark the head and foot of his grave. Nat now rests beside Kate in the Walcha Cemetery and his many admirers have clubbed together to restore his grave and record his achievements on a bronze plaque. A memorial stone has also been placed near his original gravesite at Kenmuir, by the Tamworth Historical Society.

Nat left the sum of 950 pounds in his estate. Little to show for a lifetime spent opening up the greater and harsher part of Australia to white settlement! His reward was doing what he loved best, looking for new pastoral country. In so doing he gained the respect and admiration of his peers and fellow bushmen, and left a legacy for future generations of Australians.

> *The coastal forest, the rolling downs, the semi-deserts of the great heart of Australia would miss his insistent treks, hear no more the tramp of his invading herds, and see his bearded face no longer.*

41. "Kenmuir" The farm in New England where Nat retired (circa 1900)
Buchanan Collection

42. Catherine
Buchanan Collection

Epilogue

Kate remained at Kenmuir for a couple of years, then moved to Sydney. When Flora Valley was sold, the two Gordon brothers bought a house in Chatswood and Kate and her sister Annie shared it with them. The house was named Flora Valley. Just like its namesake in the far away Kimberley it provided a welcome to visitors from the Outback. Kate continued to help the needy and she gave her energy, time and money unstintingly to many deserving causes.

In September 1924, Kate passed away quietly of a cerebral haemorrhage, and was laid to rest in the Walcha cemetery beside her husband. It had been at Kate's request that Nat's remains had been exhumed by the Nundle police and reinterred at Walcha beside Kate's parents, John and Isabella Gordon. Kate was one of those dauntless early pioneer women who helped husbands, fathers and brothers in the work of shifting the outposts further out. She was the first woman pioneer in the Thomson River district and one of the first in East Kimberley. Although much of her married life was necessarily spent apart from her husband, their mutual love, respect and understanding stood the test of time.

Wattie Gordon died in 1936, aged 83 years. The eulogies from his friends were sincere and give us some insight into the character of this old gentleman.

In 1938 at the age of 86, Hugh joined his brother Wattie for that final muster. Hugh and Wattie's names will figure largely in the history of the North, not only as pioneers but because they were respected and loved by the dwellers and travellers through that Never-Never land.

Gordie bought a farming property near Glen Innes called Malboona, where with his wife Nelly, they raised two sons - Nat and Gordon. In later years they retired to Sydney.

The depression severely depleted Gordie's finances but he maintained an active interest in the affairs of the North, politics and social issues. He wrote many articles for papers and journals on these subjects as well as some short stories. The book Gordie wrote about his father's and his own experiences in the north of Australia called *Packhorse and Waterhole* was first published by Angus & Robertson in 1933 and reprinted in 1934. In his declining years, encouraged by his friend Dame Mary Gilmore, he attempted to add to the record of his father's life by writing a manuscript called *Old Bluey*. Completed only a short time before his death in 1943, the manuscript was never published.

'Old Gordie', as he was affectionately known, was a man of simple tastes with great compassion for the working man. Though gentle and kind he was not afraid to contribute to public debate and speak out against social injustice.

43. The Gordons in retirement at their home called Flora Valley in Chatswood, N.S.W.
Standing from left - Wattie and Willie
Seated from left - Catherine and Hugh
Buchanan Collection

44. Family Outing (circa 1918)
Driver - Gordie Buchanan, seated directly behind driver Gordon Buchanan Jnr., standing on far right
Nathaniel H. Buchanan, left of front passenger Nellie Buchanan.
Buchanan Collection

Index